Understanding
the Brain

Understanding the Brain

FROM CELLS TO BEHAVIOR TO COGNITION

John E. Dowling

W. W. NORTON & COMPANY
INDEPENDENT PUBLISHERS SINCE 1923
NEW YORK · LONDON

Important Note: *Understanding the Brain* is intended to provide general information on the subject of health and well-being; it is not a substitute for medical or psychological treatment and may not be relied upon for purposes of diagnosing or treating any illness. Please seek out the care of a professional healthcare provider if you are pregnant, nursing, or experiencing symptoms of any potentially serious condition.

Portions of this material appeared in different form in John Dowling's *Creating Mind: How the Brain Works* (1998).

For information about permission to reproduce selections from this book, write to Permissions, W. W. Norton & Company, Inc., 500 Fifth Avenue, New York, NY 10110

For information about special discounts for bulk purchases, please contact W. W. Norton Special Sales at specialsales@wwnorton.com or 800-233-4830

Manufacturing by Lake Book Manufacturing, Inc.
Book design by Molly Heron
Production manager: Katelyn MacKenzie

Library of Congress Cataloging-in-Publication Data

Names: Dowling, John E., author.
Title: Understanding the brain : from cells to behavior to cognition / John E. Dowling.
Other titles: Creating mind
Description: New York : W. W. Norton & Company, [2018] | Orignally published: Creating mind / John E. Dowling. c1998. | Includes bibliographical references and index.
Identifiers: LCCN 2018009314 | ISBN 9780393712575 (hardcover)
Subjects: LCSH: Brain. | Neurosciences.
Classification: LCC QP376 .D695 2018 | DDC 612.8/2—dc23
LC record available at https://lccn.loc.gov/2018009314

W. W. Norton & Company, Inc., 500 Fifth Avenue, New York, N.Y. 10110
www.wwnorton.com

W. W. Norton & Company Ltd., 15 Carlisle Street, London W1D 3BS

1 2 3 4 5 6 7 8 9 0

For Alexander John, who has brought so much joy to our lives, and his yet-to-be-born brother

Contents

PART TWO Systems Neuroscience: Getting at Behaviors

PART THREE Cognitive Science: Higher Brain Function
and Mind

Preface

What makes us human and unique among all creatures is our brain. Perception, consciousness, memory, learning, language, and intelligence all originate in and depend on the brain. The brain provides us with wondrous things, from mathematical theories to symphonies, from automobiles and airplanes to trips to the moon. But when it goes awry, we are undone.

Over the past century, our understanding of the brain has raced forward, yet those who study the brain are still scratching the surface, so to speak. What is the mind, after all, and how does it relate to brain function? Most neuroscientists believe the mind originates in brain function, but at the moment no one can define adequately what we mean by *mind*. "Consciousness" is a particularly elusive subject, though philosophers and others endlessly talk about what it means.

As a neurobiologist, I am forever peppered with questions about the brain and brain function. This is especially true for friends who know about the exciting discoveries in the brain sciences yet witness the consequences of mental illness, aging, or brain injury and want to know more about new drug therapies for treating these problems. No field of medicine is untouched by the advances in the brain sciences, especially as we have come to realize how much the course of a disease and even its outcome, perhaps, can be affected by brain function and mental state. Sound body/sound mind is a two-way street; each is profoundly affected by the other.

This book is intended to answer many of the questions about neuroscience I am often asked. At the same time, I hope to convey to the general reader the essence and vitality of the field—the progress we are making in understanding how brains work—and to describe some of our strategies for studying brain function. Whenever possible, I try to relate topics to something relevant, such as a disease or a consequence of brain function. Much wonderful work in the field is ignored—to keep the book manageable and, I hope, an interesting read.

The first four chapters provide the nuts and bolts necessary for an up-to-date understanding of the brain. The remainder of the book dips into aspects of brain function—vision, perception, language, memory, emotion, and consciousness—seemingly more relevant to cognition and how the brain creates mind. But if an in-depth understanding of these topics is to be gained, the nuts and bolts of brain function must first be sorted out.

An earlier version of this book was published 20 years ago and was called *Creating Mind*. Where are we today in understanding brain function? Clearly we have learned much in two areas especially—first in neurobiology, the biological mechanisms underlying neuronal function. We can now describe how neurons work in exquisite detail—how they are stimulated, how they generate and propagate electrical signals along their membranes, how they transmit information at synapses, and how they may be modulated or modified on short- or long-term bases. In addition, a wealth of information has been uncovered about the glial cells in the brain—the cells that provide a supporting role to the neurons, in maintaining extracellular space, guiding developing neurons to their proper places in the brain, and pruning and modifying synapses in both the developing and adult brain.

On the other side of the coin, cognitive science has told us much about the human brain and mind and where in the brain behavioral and cognitive phenomena happen. Imaging of the brain, especially functional imaging, has become more and more sophisticated and revealing. Brain imaging has revolutionized psychology: no longer is the brain thought of as a black box.

So where do we need to go? In my view it is to integrate neurobiology and cognitive science. How do groups of neurons interact to underlie complex behaviors? This is the province of systems neuroscience. We are just at the beginning of this quest, which must happen if we are truly going to understand the brain. In my view, it is the major challenge for neuroscience in the twenty-first century. This will be important for understanding not only the normal brain but also the diseased, damaged, or cognitively compromised brain.

Invertebrates have provided some nice examples of understanding simple behaviors neurobiologically (see Chapter 5) but for complete behaviors in vertebrates we have far to go. How will this happen? New electron microscope techniques for imaging and reconstructing brain areas down to the single synapse level, recording the activity of hundreds of neurons simultaneously, and computational modeling are all being worked on, intensively and are promising in this regard. In this book, I focus on what we do know presently about how the brain functions. As a neurobiologist, I try to emphasis whenever possible underlying biological mechanisms for the phenomenon under discussion, in an attempt wherever possible to bridge the gap between neurobiology and cognition.

Chapter 1 describes the general organization of the brain. What are the cells like that are found in the brain? How do they differ from cells elsewhere in the body? Chapter 2 discusses how brain cells receive, process, and transmit information. Neural signals travel along cells electrically but between cells chemically. How do cells accomplish this? How do brain cells generate electrical signals, and how do chemical substances pass information to adjacent cells? Chapter 3 discusses in more detail how brain cells talk to one another and the changes that occur in brain cells when they are contacted by other cells. The chemical substances used to communicate signals in the brain are described, and drugs that alter chemical transmission and cause profound alterations in brain function are discussed. Chapter 4 discusses the various sensory receptor systems we have that enable us to experience the world.

Chapter 5 describes how invertebrates—animals without backbones—have been invaluable for elucidating neural and behavioral mechanisms. Animals from the sea, such as squids, horseshoe crabs, and sea slugs, have been particularly useful, and examples of important findings from these animals are presented, as well as the role of genetics in uncovering the mechanisms underlying circadian rhythmicity in fruit flies. Chapter 6 describes the architecture of the human brain—the various parts of the brain and what roles they play. How do the brains of frogs and fishes differ from our brains? Chapters 7 and 8 explore the visual system in depth, from retinal and cortical function to current ideas about visual perception. We know more about the visual system than any other brain system; it provides a wealth of clues about brain function. Chapter 9 deals with development and brain plasticity. How do embryonic brain cells find their way to their targets? How do environment and injury affect the developing and adult brain? The fascinating topics of language, memory, and learning are dealt with in Chapter 10 and 11, along with brain imaging and the question of how we discover new things about the human brain. The neurology clinic has long provided instructive examples of patients with specific brain lesions. Today, brain imaging techniques promise a wealth of new information about the human brain.

Chapter 12 turns to matters we associate more with mind—emotions and rationality. What regions of the brain are involved in emotional behaviors, and what happens when these areas are disrupted? Rationality is an example of an emergent property of the brain, deriving perhaps from emotional behaviors. Finally, Chapter 13 discusses consciousness. What do we mean by *consciousness*, and what can we say about consciousness from a neuroscience perspective? Throughout the book, I start most chapters with an example where brain function is compromised by injury or disease. Such examples are not presented simply as curiosities. Rather, these alterations in brain function cast light on normalcy. Finally, a glossary is provided to help with unfamiliar terms or concepts, but in almost every case further explanation is provided elsewhere in the book.

The earlier version of this book was written primarily for the nonscientist and it did well, having been reprinted seven times. I have used it for many years as the basic book for my freshman seminar at Harvard. These first-year students, interested in science, but not planning to major in a science, always found it accessible. After 20 years, however, the book clearly needed some updating, so I approached W. W. Norton & Company, its publisher, about doing this. I received an enthusiastic response from Deborah Malmud, a Vice President at Norton, who told me she always enjoyed the book and that it was one of the first books on her shelf when she began working in professional book publishing. She suggested that I not simply update the book, but expand it, essentially making it a new book with a significant amount of new material and with another title. This I have tried to do, describing not only what we presently know about brain function, but where we need to go—the future of brain research.

I owe many thanks to Deborah for her help throughout the writing, from her initial suggestions, to reading partial drafts and eventually the final draft. She has been most supportive. Thanks also to her assistant, Kate Prince, and project editor Mariah Eppes for their patience and enormous help. And finally, many, many thanks to Giselle Grenier, my assistant, who patiently and expertly entered the many drafts of the manuscript into the computer and also carried out all the other things such a project entails.

Understanding
the Brain

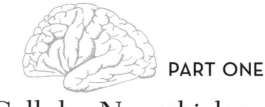

PART ONE

Cellular Neurobiology

THE NUTS AND BOLTS

Over the past century neurobiologists have focused on the cells of the nervous system. The structure of neurons and synapses has been elucidated in great detail; we now understand how individual nerve and receptor cells generate, carry, and transmit electrical and chemical signals, and we have identified many of the substances used by neurons to communicate information. More recently, molecular biological approaches are revealing the ways in which molecules carry out the various tasks of neural activity.

The Uniqueness of the Brain

Bob Jones was 62 years old when he retired as chief executive officer of a small company. He had always been an effective administrator and was dedicated to his wife and family. Two years earlier, he had become uncharacteristically short-tempered and forgetful, but this change was attributed by everyone to stress. Although retired as CEO, he continued to work for the company.

Mr. Jones became increasingly forgetful, and finally he went to see his physician, who reassured him that he was just getting a little older. His irritability increased to the point that nothing seemed right, and his wife took him back to the physician. The physician thought Mr. Jones was experiencing depression. An antidepressant drug was prescribed, along with psychotherapy, and both appeared to help for about a year.

Over the second year, Mr. Jones's memory deteriorated significantly, and he experienced attention loss and an inability to learn new things. He no longer worked, and a brain scan indicated a significant shrinkage of his brain. A diagnosis of Alzheimer's disease was made.

Mr. Jones's mental abilities now began to deteriorate dramatically. On several occasions he made a cup of coffee when his wife was out and forgot to turn the stove off. If he left the house he promptly became lost, and eventually he became lost in his own home. He became so confused about right and left that he could not put on his clothes without help. Difficulty in orienting his arms and head in space eventually made it impossible for him to eat without assistance. His wife cared for him over this time but felt increasingly that it was not her husband she was caring for but a stranger. He had lost virtually all the traits that had made him a unique individual.

    —*Adapted from David L. Rosenham and Martin E. P. Seligman, Abnormal Psychology: Casebook and Study Guide (New York, NY: Norton, 1995)*

The human brain weighs no more than three and a half pounds—only about 2–3 percent of our total body weight—but its importance cannot be overstated. It oversees virtually everything we do and makes us what we are. When the brain deteriorates, as happened to Mr. Jones, not only are individuals unable to carry out even simple tasks such as eating, but they also lose their uniqueness and individuality.

We are aware of many activities the brain controls; walking, talking, laughing, and thinking are just a few of them. The brain initiates these activities and also controls and regulates them. But as we go about our daily lives, we are unaware of many other aspects of brain function: the regulation of internal organs, including the heart and vascular system, lungs and respiratory system, and gut and digestive system. The brain also coordinates and integrates movements employing mechanisms that we don't notice, such as the use of abundant sensory information from muscles, tendons, and joints. The extent of muscle contraction is signaled to the brain, yet ordinarily we are quite unaware of the state of our individual muscles.

Of most interest, and most mysterious, are mental functions referred to as "mind." Feelings, emotions, awareness, understanding, and creativity are well-known aspects of mind. Are they created in and by the brain? The consensus today among neuroscientists and philosophers is that mind is an emergent property of brain function. That is, what we refer to as *mind* is a natural consequence of complex and higher neural processing. Clearly, brain injury or disease can severely compromise the mind, as happened to Mr. Jones. At the very least, then, mind depends on intact and healthy brain function.

Do animals have minds? The answer to this question depends mainly on one's definition of *mind*. Certainly cats, dogs, and monkeys can express emotion, show some understanding, and even apply creative approaches to simple problems, but no animal approaches humans in richness of mind. Is there something unique about human brains relative to those of other organisms? Not that we know. So how do we explain our extraordinary mental abilities?

Is it that human brains represent a higher evolutionary level than the brains of other animals? This is likely so, even though the cellular mechanisms underlying human brain function are similar to those operating in other animals, even those that have very elementary nervous systems and exhibit virtually no aspects of mind. The view I espouse

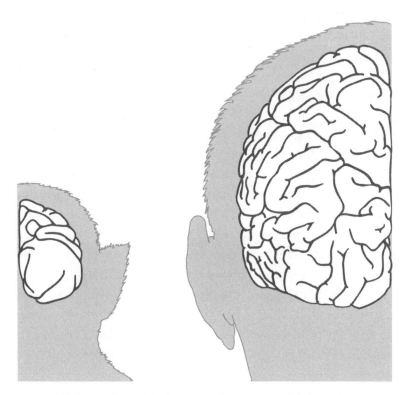

FIGURE 1.1 *The brain of an adult rhesus monkey compared to that of a human. Not only is the human brain very much larger, but its surface also has many more folds, thus increasing the surface (cortical) area of the brain substantially.*

is that the human brain is qualitatively similar to the brains of other animals but quantitatively different. That is, the human brain has more nerve cells than do the brains of other primates, our closest relatives, but also, and what is probably more important, the cerebral cortex of the human brain, the seat of higher neural function—perception, memory, language, and intelligence—is far more developed than is the cerebral cortex of any other vertebrate. (Figure 1.1) And because of the added neural cells and cortical development in humans new facets of brain function emerge.

An understanding of the brain requires knowing its structure, function, and chemistry. Once we do so, can we truly understand the mind? Can the brain understand itself? No one knows; there is much to learn. To lay the groundwork for this quest, I first describe the brain's elements and their structure, how the elements communicate with one another, their special features, and the consequences of these features. This will bring us to a general notion of how the brain is organized.

Cells of the Brain: Neurons and Glia

The brain, like other organs, is made up of discrete units or cells. Two classes of cell make up the brain: *neurons*, or nerve cells, whose business is to receive, integrate, and transmit information; and *glia* (derived from the Greek word for glue), which are supporting cells. Glial cells do things like maintaining the neurons and the brain's environment, even pruning neuronal branches and terminals. They regulate the levels of substances needed or used by neurons in the spaces between the cells. They also provide a structural framework for neurons (especially during development), and they insulate the neurons to make them conduct electrical signals more effectively. But the key cells for understanding how the brain works are the neurons, and the brain contains billions of them.

We don't know exactly how many neurons are in the human brain, but the best estimates suggest about 86 billion—ten times more than a chimpanzee (7 billion). Neurons are also elaborate and have numerous extensions or branches that may extend long distances. For example, a neuron that controls muscles in the foot has a branch, called an *axon*, that extends down the length of the leg to the foot, a distance of about 3 feet (or a meter). The body of this cell resides in the lower part of the spinal cord and is less than 0.1 millimeter in diameter. To put the difference between the size of the cell body and the length of its axon in perspective, consider how long the axon would be if the cell body were 6 inches wide—the axon would extend almost a mile!

The branches of neurons allow them to contact one another in the brain in complex and intricate ways. Typically, neurons make one hundred to ten thousand connections with other neurons, and one type of neuron (the cerebellar *Purkinje cell*) makes as many as 100,000 connections.

Neurons have many branches, so most of the brain consists of neuronal branches. Two kinds of neuronal branches are distinguished anatomically: *dendrites* and *axons*. Dendrites are like the branches of a tree, relatively thick as they emerge from the cell but dividing often and becoming thinner at each branch point. Many dendrites usually extend from each neuron. Axons, in contrast, are thinner at their point of origin on a neuron and remain constant in diameter along most of their length. Neurons usually have just one axon that branches profusely as it terminates. Input to nerve cells occurs on the dendrites and cell body; the axon carries the cell's output.

Figure 1.2 depicts a neuron with a short axon found in the retina of the eye. Each tiny branch in both the dendritic tree and *axon terminal* complex probably represents a point of functional contact with another cell. And it is likely that a number of contact points were missed by the scientist who drew the cell. The functional contacts between neurons, called *synapses*, operate mainly chemically. That is, at synapses neurons release specific chemical substances that diffuse to an adjacent neuron contacted by that synapse. This chemical may excite, inhibit, or modulate the contacted neuron.

Synapses have a characteristic structure that can be readily visualized with the high resolving power of an electron microscope. Figure 1.3 (top) shows a drawing of a synapse (essentially one of the axon terminal branches in Figure 1.2). Below to the right is an electron micrograph of a synapse, and a drawing of it is on the left. What you are viewing in the electron micrograph is a slice through the middle of the axon terminal that is shown in the drawing above. A prominent feature of axon ter-

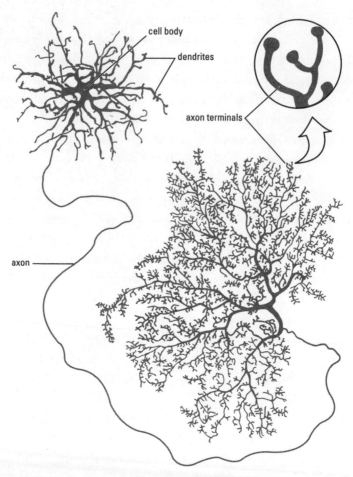

FIGURE 1.2 *A neuron with a relatively short axon. Input to a neuron is usually onto the dendrites or cell body, whereas the cell's output occurs via the axon terminals. Each tiny branch in both the dendritic and axonal terminal arbors is likely to be a site of synaptic contact.*

minals is the presence of tiny *synaptic vesicles* within the (*presynaptic*) terminal: the vesicles store the chemicals released at the synapse. The vesicles cluster at the synapse, fuse to the cell *membrane* when synapses are active, and release their contents into the small space between the two cells (arrow in Figure 1.3). The chemicals spread across the space and interact with specific molecules (*proteins*) present within the membrane of the contacted (*postsynaptic*) cell. The molecules, when activated by the synaptic chemical, initiate mechanisms that alter the postsynaptic cell.

Neuroscientists focus much attention on synapses because of the general agreement that synaptic interactions between neurons explain much of what the brain does. For example, most drugs that affect states of behavior, such as cocaine, LSD, Prozac, and even Valium, do so by modifying synaptic activity (see Chapter 3). Furthermore, affective mental disorders, such as schizophrenia, depression, and anxiety, appear to result from impaired synaptic mechanisms in the brain.

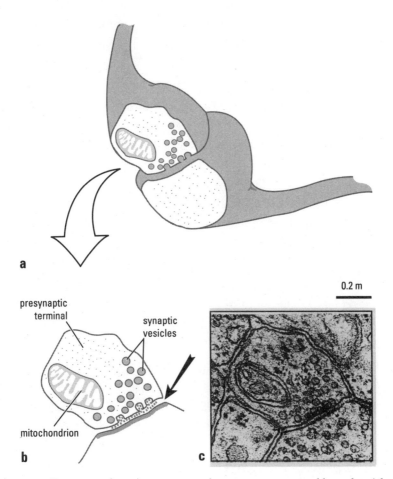

FIGURE 1.3 *Drawings of a synaptic contact between two neuronal branches (above and below to the left) and an electron micrograph of a synapse (below right). The synaptic vesicles contain substances that are released when the synapse is active. The vesicles attach to the membrane and release their contents into the small space between the two branches (arrow). The substances released can excite, inhibit, or modulate activity in the contacted branch.*

How Special Are Neurons?

Neurons employ the same cellular mechanisms as do other cells in the body. Each neuron has a *nucleus* containing *nucleic acid (DNA)* that specifies the proteins made by the cell. Neurons possess *ribosomes*, structures responsible for assembling proteins, and *mitochondria*, which supply energy-rich molecules that power the cells. Neurons also contain tubules and filaments found in virtually all cells; these are involved in the movement of substances throughout the cell, and they help maintain the complex structure of a neuron.

It is clear from this that the neurons' biochemical mechanisms are similar to those used by all cells. Yet neurons differ from other cells in the body in two significant ways that have important medical consequences: first, most brain cells are not replaced when lost, and second, brain cells constantly require oxygen. Once neurons have matured during embryonic development, they can never divide again. This is quite different from most cells in the body that dedifferentiate, divide, and produce new cells in response to injury or disease. A cut on the finger soon heals as cells divide and fill in the injury. The same is true for cells in most organs of the body.

When brain cells are lost because of injury or disease, they are not usually replaced. The brain of a 1-year-old human contains almost as many cells as it will ever have, and throughout life neurons are lost via normal aging processes. In other organs, dead cells are quickly replaced, but in the brain they are not. And the number of brain cells lost can be surprisingly high—maybe as many as 150,000 per day. This estimate comes from the finding that, with age, at least 5–10 percent of brain tissue is lost. If you assume a loss of 7 percent of the brain cells in eighty to 100 years, with, say, 86 billion cells at the outset, about 150,000 cells are lost per day. With age, the brain also loses neuronal branches and the neurons shrink in size, and this also causes loss of brain volume.

Cells in most tissues of the body are replaced in two ways: mature cells may dedifferentiate and then proliferate to form new cells for that tissue, or *stem cells* may be present in the tissue, which are capable of dividing and forming new cells. However, in the brain, stem cells are found prominently only in two areas; in the hippocampus, which is involved in long-term memory formation, and in the olfactory system. And it appears that in humans there are no stem cells in the olfactory system. Further, it is not clear what role the stem cells play in the hippocampus. The neurons they produce do not live that long, whereas the other cells in the brain live as long as a human lives, 80–100 years. Further, the generation of new neurons decreases with age, and in older humans, new neurons have been difficult to identify in the hippocampus.

If neurons cannot divide once they have matured and there are few stem cells throughout the brain, how can a brain tumor grow? In adults, most if not all brain tumors are glial cell tumors. Glial cells, unlike neurons, can divide in the adult brain, and when glial cell division becomes uncontrolled, a tumor or cancer can result. Only in children do brain tumors arise from neurons, and fortunately these are quite rare.

Because the brain contains so many neurons, most of us can get through life without losing so many cells that we become mentally debilitated. Eventually, though, brain cell loss with age does catch up with us, and eventually mental deterioration takes place in virtually everyone. It's a mystery why some individuals maintain keen mental abilities much longer than do others; indeed, it may be brain cell loss that determines human life span. If we could eliminate heart attacks, cancer, and other fatal diseases, we still might not extend the absolute life span of humans, because most brain cells cannot divide and replace themselves. Although the average life expectancy for humans has increased more than 60 percent since 1900, from about 50 years in 1900 to over 80 years today, the maximum number of years that humans live has not increased appreciably since ancient times. Figure 1.4 shows the trends in human longevity from antiquity to the present day. Medical advances and improved housing and sanitation have greatly increased the numbers of people who live to age sixty, from about 20 percent of the population to nearly 80 percent. But only a very small percentage of humans live to be over 100 years of age. The oldest well-documented individual to die was 122 years of age, and that was more than 20 years ago. No one has lived as many as 120 years since.

Alzheimer's disease is marked by excessive brain cell loss. As many as four million Americans probably suffer from the disease, and it is

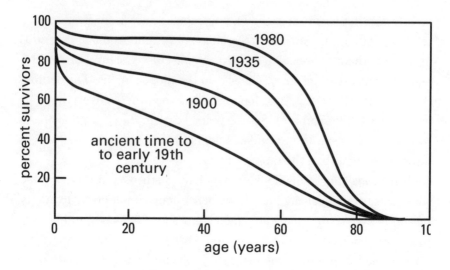

FIGURE 1.4 *A graph showing the changes in life expectancy and life span of humans from ancient times to 1980. Although life expectancy has changed dramatically from an average of 35 years 200 years ago to over 75 years in 1980, the absolute life span of humans has not increased significantly since antiquity. That is, only a very small percentage of humans live beyond 100 years of age.*

estimated that as many as fourteen million may be affected by the year 2040. Those who suffer from it may have a decline in their mental abilities (dementia) in their late fifties or early sixties, beginning with deficits in recent memory and progressing to a loss of virtually all higher mental functions. The symptoms displayed by Bob Jones described at the beginning of this chapter are typical. Confusion and forgetfulness, followed by sharply declining motor abilities and even loss of speech, are common symptoms. Whether the brain cell loss in Alzheimer's disease is due to an acceleration of normal brain cell loss or the result of a disease is still unclear, although most neurologists believe the latter is so. By the age of sixty-five, about 10 percent of the population show Alzheimer-like symptoms, and by age eighty-five as many as half the population may have some symptoms.

Severe brain injury or trauma also leads to excessive brain cell loss, and these individuals often display Alzheimer-like symptoms. People prone to brain injury, such as prizefighters, have these symptoms, and such individuals are sometimes referred to as "punch-drunk." Brain cell death may not be evenly distributed throughout the brain, which leads to different symptoms for people with Alzheimer's disease, brain injury, or normal aging. The former heavyweight boxing champion Muhammad Ali had Parkinson's disease, a degenerative brain disease affecting parts of the brain's motor control system. He walked sluggishly, had difficulty initiating movements, and displayed an expressionless face, all classic symptoms of Parkinson's disease. It is likely that his disease was linked to the pounding his head took during his many years as a prizefighter. Recently, it has been shown that in contact sports, such as American football and ice hockey, brain cells can be injured, resulting in excessive brain cell death, and this can result in dementia and other neurodegenerative afflictions. Often such individuals have suffered several concussions during their playing days, and this is a sign of impending brain damage.

Alzheimer's disease is probably not a single entity but several diseases. For example, early-onset Alzheimer's disease is clearly inherited in a number of instances, but in most cases (about 90 percent) the disease is probably not directly inherited. Certainly some individuals have a genetic predisposition for Alzheimer's disease, but what factors cause some individuals such as Bob Jones to develop the disease in their sixties and others never, or very much later in life, are not understood.

The second critical difference between neurons and cells from other tissues is that brain cells constantly require oxygen. When deprived of it, neurons die within minutes. Other cells can survive without oxygen and even function without oxygen (anaerobically) for some time. During a hundred-yard race, a runner's leg muscle cells use up the available oxygen after only about thirty yards. The muscles keep functioning without oxygen by chemically breaking down sugar and producing energy-rich molecules by a fermentation-like process. After the race, as oxygen is restored to the muscles, the cells break down these sugar fragments and restore the muscle to its resting state. The oxygen required to repay the so-called oxygen debt is provided after the race when the runner breathes hard for several minutes. If your muscles are sore after heavy exertion, the reason is partly because of an excessive buildup of the breakdown products of sugar, especially lactic acid, during the anaerobic phase of muscle activity.

Neurons, however, cannot survive anaerobically, even for a short time. After a person has a heart attack or suffocates, oxygen flow to the body's tissues is shut off and the brain will quickly die. If oxygen is restored in just a few minutes, the brain can survive, but time is of the essence. It is not uncommon for a patient suffering a severe heart attack to be left permanently brain-dead after a short period of oxygen deprivation, whereas today, with effective life-support systems, other organs like the heart and kidneys survive and recover completely.

The brain's need for oxygen is so acute that, when a part of the brain is active, blood flow rapidly increases in that region. This is the basis for *brain imaging* techniques like *positron emission tomography (PET)* and *functional magnetic resonance imaging (fMRI)*, which enable neuroscientists and physicians to probe the brains of awake subjects, when they are involved in specific behaviors (see Chapter 10 and Figure 10.5). These powerful techniques provide neuroscientists and physicians with a wealth of information about the localization of brain phenomena and with critical clinical information in brain disease.

When a person suffers a stroke, blood flow to the brain is interrupted, and the region losing its blood supply usually dies because of the lack of oxygen and loss of nutrients. Although an immediate deficit is evident in most stroke patients, everyone who has observed a surviving stroke patient knows that some recovery occurs, and it can continue for months, long after the first acute changes, such as swelling of the brain, have subsided. The recovery is sometimes virtually complete. The same can happen after a severe brain injury; remarkable recovery sometimes takes place.

If brain cells die after a person has a stroke or injury, and if they are not replaced, how does one recover, especially over the long term? The answer is that remaining brain cells are used; that is, nearby cells can take over for the damaged or dead brain cells, which enables at least a partial recovery. How extensive the recovery is depends on the extent of the damage and on the region damaged. When some parts of the brain are damaged, little or no recovery is seen, but many parts are more forgiving or plastic. Other neurons can take over for lost or damaged cells mainly by sprouting new processes and forming new synapses. The mechanisms underlying for this are probably similar to those that occur in the developing brain and during the formation of long-term memories (see Chapters 9 and 11).

Brain Organization

The brain is far from homogeneous. It consists of many parts, each concerned with a separate facet of neural function. Furthermore, each part of the brain has quite distinct neuronal shapes. These differences in structure presumably relate to the role of the cells and that part of the brain. Figure 1.5 shows three cells, two from one region of the brain, the other from a second brain region. The enormous dissimilarities in cell structures are readily apparent. We don't yet understand why neurons

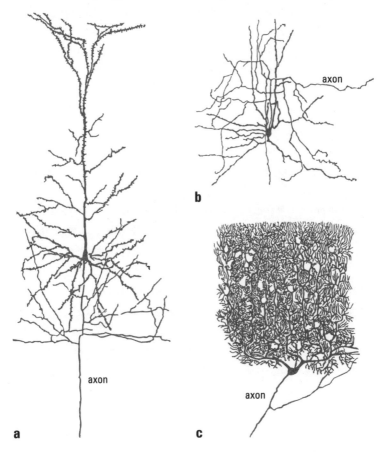

FIGURE 1.5 *Three types of neurons found in the brain. The pyramidal (a) and stellate (b) cells are present in the cerebral cortex, whereas the Purkinje cell (c) is found in the cerebellum. These drawings illustrate the enormous diversity of neuronal structure found in the brain.*

have the shapes they do, and finding out the reasons remains an enticing challenge for neuroscientists.

The distinct types of neurons are easily recognizable across individuals and even across species. A *pyramidal cell* (Figure 1.5a) in the cortex of the human brain looks similar to a pyramidal cell in the rabbit's cortex. Likewise, a Purkinje cell (named after its discoverer, Jan Purkinje, a Czech whom many regard as the founder of *histology*, the study of tissues) in the monkey looks like a cat's Purkinje cell (Figure 1.5c). Pyramidal cells are easily distinguished from Purkinje cells and would never be confused, regardless of the animal brain in which the cell is observed.

How many different kinds of cells are there in a particular brain region? In any one part of the brain, there are usually only a few major cell classes. The cerebellum has five major cell classes, the retina five, and the cerebral cortex perhaps just two. All cells of a specific class look roughly similar to other cells of that class and may play a somewhat similar functional role. But in most cases anatomists have divided the major cell classes into subtypes, of which ten to twenty or more can exist in a given region. Physiologically the subtypes usually respond in somewhat different ways, so the anatomical differences do have functional meaning.

Recent genetic studies, on the other hand, have shown that there is extraordinary diversity among neuronal subtypes, so that it is possible that no two neurons are alike genomically. In other words, based on genetics, some investigators suggest there may be as many as 86 billion neuronal subtypes, or as many subtypes as there are neurons in the brain. Thus, how to classify neurons in a meaningful way remains an open question.

It is convenient, however, to divide the various cell classes in the brain into two groups: long-axon cells and short-axon cells. Long-axon cells carry information from one part of the brain to another; short-axon cells are confined to a single part of the brain (Figure 1.5b). Short-axon cells participate in local interactions between neurons and for this reason are often called *association neurons*; they are typically involved in integrating and processing information. The greatest of the early brain histologists, Santiago Ramón y Cajal of Spain, who carried out his brain studies from the early 1880s to the 1930s, pointed out that brains of more highly developed animals contain relatively more short-axon cells than long-axon cells. This is indicative of the role that short-axon cells play in complex brain processing.

Throughout most parts of the brain, neurons are clustered together to form structures called nuclei. A brain *nucleus* usually carries out a specific neural task; for instance, a nucleus in one part of the brain regulates heart rate, while another nucleus in the same part of the brain controls respiration. Nuclei in another part of the brain regulate body temperature, hunger, or thirst, whereas nuclei in yet other parts of the brain are involved in the initiation of movements or in the transmission of specific sensory information from lower brain centers to higher ones. The general organization of a nucleus is illustrated in Figure 1.6. Information arrives at a nucleus via the axons of long-axon cells. The terminals of

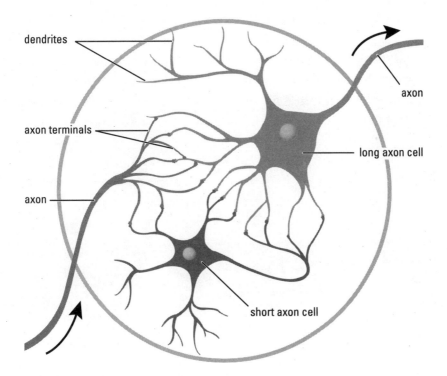

FIGURE 1.6 *A schematic drawing of a brain nucleus. Information enters and exits (arrows) the nucleus via the axons of the larger, long-axon cells. Smaller short-axon cells mediate synaptic interactions within the nucleus—their branches are confined to the nucleus.*

these cells synapse onto the dendrites of long-axon and short-axon cells. The branches of the short-axon cells are typically confined to the nucleus and make all of their synapses within the nucleus, on long-axon cells or other short-axon cells. Information leaves the nucleus via the axons of long-axon cells.

In some regions of the brain neurons are arranged in continuous layers rather than in discrete nuclei. (Figure 1.7) But the same general principles of organization persist. That is, the short-axon neurons are restricted to a relatively local area, whereas long-axon neurons carry information from one layer to another. The cerebral cortex is organized this way, as is the retina. However, it is also the case that the cortex is divided into areas that carry out specific tasks. For example, area V4 (V for visual), found in the part of the cortex concerned with vision, primarily processes form information, whereas an adjacent area, V5, is concerned primarily with analyzing moving stimuli. Such cortical areas are equivalent in functional terms to brain nuclei but are not as well defined anatomically. In other words, it is often difficult to decide where one cortical area ends and another begins.

An individual nucleus or a specific area of the brain is thus charged with a certain neural task, but the tasks are not equivalent. Some nuclei and brain areas process or regulate a basic neural function, but other

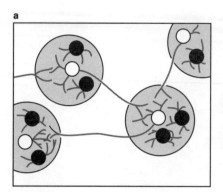
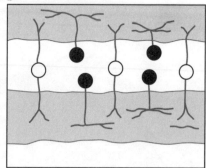

FIGURE 1.7 *Common arrangements of neurons in the brain. In many regions, neurons are clustered in nuclei (a). In other regions, neurons are arranged in layers (b). The short axon neurons (filled) are confined to a nucleus or synaptic layer (shaded areas).*

nuclei and brain areas tackle higher or more subtle aspects of processing. For example, severe damage to the primary *visual area*, V1, brings about complete and permanent loss of visual perception related to that part of the visual field served by that portion of the cortex. Yet similar damage to area V5 may lead primarily to a loss of movement perception; other aspects of vision are maintained. And damage to yet higher visual areas might lead only to difficulty in recognizing specific objects, for example, faces.

The consequence of this hierarchical organization is that tiny lesions in some parts of the brain can have devastating effects while larger lesions elsewhere can have minimal effects. Furthermore, after damage to certain areas or nuclei, little if any recovery happens, but damage to other areas may cause little if any long-term deficit. The bullet that killed John F. Kennedy entered a part of the brain where there are many nuclei that regulate vital bodily functions, such as respiration and heart rate. This region is also, from an evolutionary standpoint, an older part of the brain and more inflexible. Scant recovery happens when this part of the brain is damaged. A similar bullet wound in the neocortex—a more recently evolved brain region concerned with higher mental processing—might have caused slight permanent damage, and President Kennedy could have survived and lived a long life.

Neuroscientists are trying to discover why recovery from injury such as a stroke to some areas is much more complete than that from injury to other areas. A generalization emerging is that more recently evolved parts of the brain, those concerned with higher neural functioning, are more flexible or plastic than are older parts of the brain. This plasticity is reflected in the ability of a part of the brain to reorganize itself after damage and to recover function (see Chapter 9). Stated in cellular terms, neurons in some parts of the brain are more capable of extending new branches and forming new synapses than are neurons in other parts of the brain. The brain's structure, then, is very heterogeneous, made up of many distinct areas concerned with different neural functions. Furthermore, the neurons found in various brain regions are often anatomically distinct, and these structural differences are thought to relate to the role of a specific brain area.

Development of the Brain

The brain begins to form in humans about 3 weeks after conception. A group of cells, about 125,000 in number, form a flat sheet along the dorsal (back) side of the embryo. All of the neurons and glial cells of the nervous system derive from these cells, known as the *neural plate*.

Between the third and fourth week of development, the neural plate folds inward and creates a groove that eventually closes into a long tube, the *neural tube*. All of what is called the *central nervous system* derives from the neural tube; the anterior part becomes the brain proper, and the posterior part becomes the spinal cord. By about forty days of develop-ment, three swellings along the anterior part of the neural tube become apparent; these eventually form three areas—the forebrain, midbrain, and hindbrain. As the neural tube forms, some cells on either side are left behind.These cells, known as *neural crest* cells, come to lie on either side of the neural tube. Much of the *peripheral nervous system* derives from the neural crest cells—those nerve and glial cells that lie outside the brain and spinal cord. Figure 1.8 presents a schematic view of how the neural tube and neural crest form.

What causes the formation of the neural plate? Covering the early embryo are cells of the *ectoderm*, which eventually create skin. Lining the inside of the embryo are cells of the *endoderm*, which will form the stomach, intestine, and other internal organs. At about 2.5 weeks of devel-opment, a third, intermediate layer of cells forms, the *mesoderm*. The mesoderm differentiates into many tissues, including muscles, skeleton, and cardiovascular system; it turns out that the mesoderm is responsible for generating the neural plate cells. As the mesoderm forms, it slides under the ectoderm. As it slides under ectodermal cells on the dorsal sur-face of the embryo, it induces these cells to change their fate—to become neural plate cells rather than skin (Figure 1.8).

Experiments with salamanders first demonstrated that neural plate cells are induced from ectodermal cells by underlying mesodermal cells. In the 1920s the German biologist Hans Spemann and his student Hilde Mangold showed that if clumps of mesodermal cells are transplanted from one part of the embryo to another, the transplanted cells will induce overlying ectodermal cells to become neural plate cells regardless of where on the embryo surface those cells reside. So, for example, a second neural plate can be induced to form on the ventral side of the embryo if the transplant is placed there. These seminal experiments evoked much interest and it was further shown that if mesodermal cells are prevented from migrating under the dorsal ectodermal, no neural plate develops and the embryo fails to form a nervous system. This finding confirmed

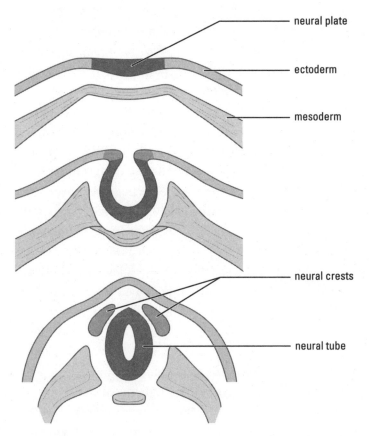

neural plate

ectoderm

mesoderm

neural crests

neural tube

FIGURE 1.8 *Formation of the neural tube and neural crest cells from the neural plate. The neural plate cells are induced from ectodermal cells by chemical signals released from underlying mesodermal cells. During the third and fourth weeks of development of the human embryo, the neural plate invaginates, forming the neural tube. The neural tube differentiates into the central nervous system—the brain and spinal cord. The neural crest cells are derived from neural plate cells positioned laterally along the plate and left behind during the formation of the neural tube. Neural crest cells differentiate into the peripheral nervous system, as well as other cells related to neural system function.*

the notion put forward by Spemann and Mangold that the neural plate cells are induced by mesoderm. Hilda Mangold was tragically killed in a kitchen explosion at the time the first experiments were published. Spemann, whose brilliant career in developmental biology extended from the turn of the century to the end of the 1930s, was awarded a Nobel Prize in 1935, primarily for his work on induction.

How can mesodermal cells induce ectodermal cells to become neural plate cells? An early and obvious suggestion was that mesodermal cells release chemical substances that induce ectodermal cells to change their fate. Several experiments support this idea. For example, when pieces of ectoderm are cultured in the presence of mesoderm, they become neural plate cells—but not when mesodermal cells are absent.

Figure 1.9 shows the development of the human brain from the neural tube. The detailed drawings on the top of the figure are enlarged relative to the middle and bottom drawings. By 60 days after conception, various brain regions can be readily distinguished. Infolding or wrinkling of the brain's surface, to increase cortical area, begins at about 7 months. By 9 months of gestation, the brain overall looks quite adult-like, but it has far to go. The average newborn human brain weighs less than 400 grams, whereas the typical adult human brain weighs about 1,400 grams. Much of the weight increase takes place during the first 3–5 years after birth, but the brain does not reach maximum weight until about 20 years of age (Figure 1.10b). Thereafter, brain weight and volume decline slowly but steadily.

Virtually all neurons are generated in the developing brain during gestation, indeed, mainly in the first 4 months of gestation. What, then, underlies the enormous growth of the brain in the first 3–5 years of life? Several things are going on, including an increase in the number of glial cells and growth of blood vessels. The most important factor, however, is the development and elaboration of the neurons themselves. Their cell bodies increase in size as they extend new dendritic branches and form new synapses. (Figure 1.10a) Probably more than 80 percent of dendritic growth occurs after birth, with a concomitant increase of synapses. But the situation is more complex than just adding dendrites and synapses. There is during early development an over production of dendrites and synapses, resulting in a substantial rearrangement and pruning of den-

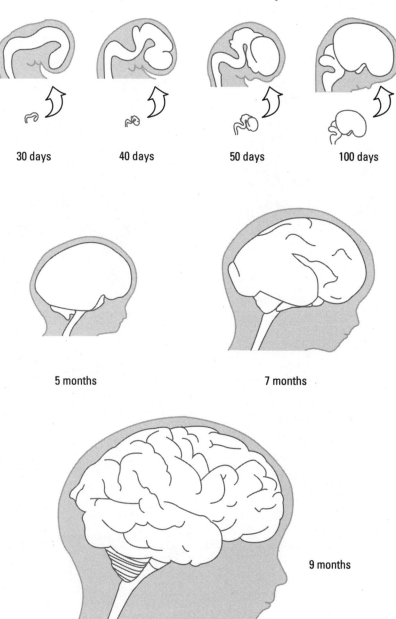

30 days   40 days   50 days   100 days

5 months   7 months

9 months

FIGURE 1.9 *Development of the human brain from the neural tube. The detailed drawings at the top are enlarged relative to the middle and bottom drawings. The smaller drawing at the top indicate roughly the actual sizes of the developing brain at early stages relative to those at latter stages roughly. Note that the infolding of the brain's surface occurs rather late in development.*

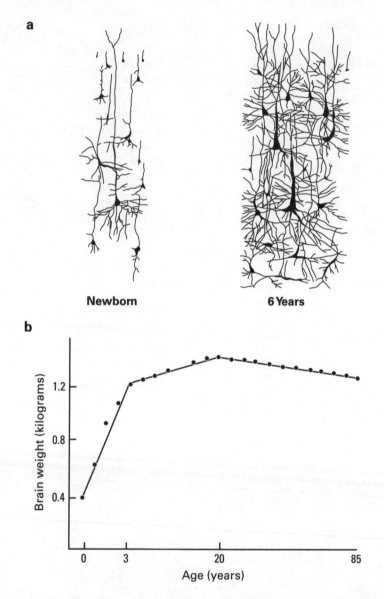

FIGURE 1.10 (a) The elaboration of neurons during brain maturation. The cell bodies increase in size along with an enormous development of the number, extent, and complexity of the cell's branches. (b) Brain weight as a function of age. A rapid increase in brain weight occurs in the first 3 years. The rate of increase then slows, but the brain does not reach its maximum weight until about 20 years. Thereafter, there is a slow and constant loss of brain weight.

drites and synapses, as brain circuits are refined and mature. Much of the pruning of processes and synapses appears to be managed by glial cells. Experience clearly influences the rewiring of brain synapses in the developing brain, but rewiring is not limited to the developing brain—it occurs to some extend throughout life, as I discuss later.

Chapter 2 explores how individual neurons work. Surprisingly, perhaps, all neurons function in basically similar ways. Understanding how one neuron works tells us pretty much how all neurons work. This is not to say that all brain neurons function identically, but that the general principles outlined in Chapter 2 hold for all neurons.

2

Brain Signals

....

The kidneys cleanse the blood, ridding the body of waste products. In a single day, the blood is washed about ten times. Kidney failure results in a condition called uremia because of a dramatic increase in urea levels in the blood. Urea is the chief waste product of nitrogen metabolism in the body and is the major dissolved substance in urine. In acute kidney failure, urea levels can increase in the blood by as much as twentyfold to reach levels of up to 2 grams per liter—a truly prodigious amount!

Patients with uremia exhibit a variety of symptoms, many of which reflect upsets in brain, neural, and muscle mechanisms. Severe headaches occur early, along with muscle twitching in the limbs. Visual disturbances are frequently reported as well. Convulsions and coma occur in the later stages of uremia, with death ensuing, usually because of fibrillation of the heart—uncoordinated and uncontrolled twitching of the heart muscle.

What causes these neurological and cardiac catastrophes in uremia? Not the large increases in blood urea levels, as one might think, but much more modest increases in one positively charged ion: *potassium* (K^+). In kidney failure, K^+ ion levels go up from about 0.3 gram per liter of blood to about 0.8 gram per liter, but this change is sufficient to cause death and the neurological symptoms described above. Death by lethal injection, a method of execution, results from raising blood K^+ levels by the direct injection of K$^+$ into the bloodstream. Understanding the key role of K$^+$ (and other ions) in generating and regulating the electrical activity of the brain, nerve, and muscle cells has been one of the triumphs of twentieth-century biology.

During the last half of the last century, neuroscientists made astonishing progress in understanding how individual neurons function. The structures of neurons and synapses was elucidated in great detail. We understand how single nerve cells are excited by sensory stimuli such as light or sound, or synaptic input, and how neurons carry and code information. To understand how neurons operate, we need to look beneath the surface and see in some detail how single neurons process and carry information and how neurons communicate with one another. Communication between neurons (at synapses) is mainly chemical, whereas individual neurons usually code and carry information electrically. This chapter describes how the electrical and chemical signals are generated, how they encode information, and what happens when the signals are interrupted. To start, I provide a short primer on electricity.

Electricity and the Brain

Electrical charge is an intrinsic property of all matter. It comes in two polarities, positive and negative. Charges of the same polarity repel; charges of the opposite polarity attract. Two of the fundamental particles that make up atoms, *electrons* and *protons*, have charge. Electrons possess negative charge, whereas protons possess positive charge. The electricity used at home consists of electrons moving through metal wires. Electrical *current* is a measure of how many electrons move through a conducting medium, such as metal wire, during a period of time.

Atoms ordinarily have equal numbers of electrons and protons and are therefore electrically neutral. But atoms can gain or lose electrons and become charged, either negatively or positively (Figure 2.1a). Charged atoms are called *ions*; they generate the electrical signals in the brain when they move in and out of cells. Most ions have one extra positive or negative charge, but some ions important for neural function have two extra positive charges. A cell's outer membrane does not readily allow

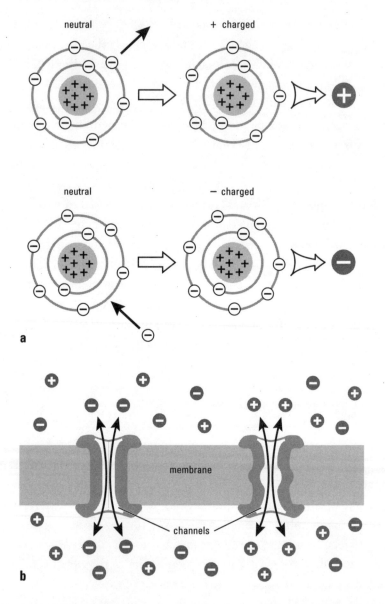

FIGURE 2.1 (a) Electrically neutral atoms become positively charged ions (top) or negatively charged ions (bottom) by losing or gaining an electron. A positively charged ion has more positively charged protons in the nucleus than negatively charged electrons surrounding the nucleus. A negatively charged ion has more electrons than protons. (b) A cell membrane is impermeable to ions. However, protein channels that span the membrane allow ions to flow from one side of the membrane to the other. Some channels allow only negatively charged ions to cross the membrane (left); others, only positively charged ions (right).

ions to cross, but proteins in the membrane can form *channels* that allow and regulate the flow of ions across the membrane (Figure 2.1b). These channels can often discriminate among ions of different charge (such as sodium, a positively charged ion, versus chloride, a negatively charged ion) and even among different ions carrying the same charge (such as sodium versus potassium). Today, we have some notion of how protein channels can discriminate ions of similar charge and how channels open and close. Some channels are open all the time, but most channels allow ions to cross the membrane only in response to a stimulus, such as a chemical released at a synapse.

To keep ions flowing through the channels when they are open, cells maintain different concentrations of ions on either side of the membrane. They accomplish this with ion *pumps*: specific membrane proteins that actively transport ions from one side of the cell membrane to the other. The pumps require energy and are generally always operating, moving a few ions at a time across the membrane. When a channel opens, ions move through it in an attempt to equalize ion concentrations on both sides of the membrane. When ions cross the membrane, they alter the charge across the membrane. Membrane *voltage* (or *potential*) is a measure of the charge difference across a membrane. Voltages of 10–100 millivolts (0.01–0.1 volt) typically occur across cell membranes when a cell generates a signal. These are tiny voltages compared to the 110 volts powering our home electrical appliances.

Cell Resting Potentials

In addition to the electrical signals generated when nerve cells are stimulated, a resting voltage (also called the *resting potential*) exists across the outer membrane of all cells. In nerve cells, this resting voltage is about 70 millivolts (0.07 volt). The inside of the nerve cell is negatively charged relative to the outside; there are more negatively charged ions inside the cell than positively charged ions, whereas outside the cell there is an excess of positively charged ions. This charge imbalance results in the resting voltage that is measured between the inside and outside of the cell. This happens because opposite charges attract; thus, a force exists across the relatively impermeable cell membrane. The greater the charge imbalance between inside and out, the greater the force; that is, the greater the voltage or potential. Understanding how resting potentials come about can help us understand how other electrical signals are produced.

Two factors underlie the resting potential of cells. First, ions are

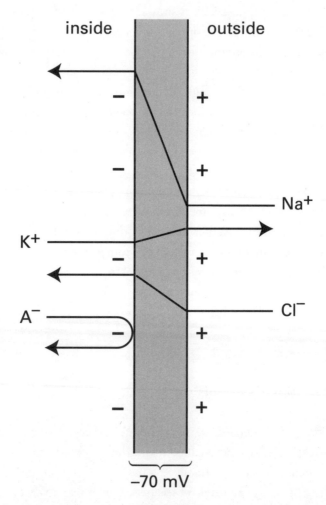

FIGURE 2.2 A schematic rendering of the location of the principal ions inside and outside a nerve cell and the relative permeability of the cell's membrane to the ions. K^+ and organic anions (A^-) are the major ions inside nerve cells; Na^+ and Cl^- predominate outside nerve cells. The membrane is more permeable to K^+ than to Cl^- or Na^+ and virtually impermeable to A^- at rest. This results in a resting potential of about -70mV across the cell membrane.

unevenly distributed on the two sides of a cell's membrane as a result of the action of the ion pumps present in the membrane. Four ions are principally involved: *sodium (Na⁺), potassium (K⁺), chloride (Cl⁻),* and small organic molecules that have an excess negative charge and act like ions (A⁻) (Figure 2.2). Potassium ions (K⁺) and the small negatively charged organic molecules (A⁻) are present predominantly inside cells, whereas Na⁺ and Cl⁻ are mainly outside cells. Second, the membrane is differentially permeable to these ions. It is most leaky to K⁺, not at all to A⁻, and has low to moderate permeability to Na⁺ and Cl⁻, respectively, as shown in Figure 2.2 by the steepness of the line—almost flat for K⁺, very steep for Na⁺. This means that nerve cell membranes in the resting state have channels that allow K⁺ to pass through relatively easily but no channels that allow A⁻ to pass through. Further, the channels that allow Na⁺ to pass through the membrane are virtually closed at rest.

How a difference in ion concentrations between the inside and outside of a cell, coupled with a differential permeability of the cell's membrane to these ions, results in a resting voltage can be illustrated by a simple model of a cell that involves just K⁺ and A⁻ (Figure 2.3). As in a real cell, we make the concentrations of K⁺ and A⁻ high inside our model cell and low outside (indicated by size of letters K + A⁻). Furthermore, we make the membrane permeable to K⁺ but not to A⁻. In other words, the channels in the membrane allow K⁺ to pass through but not A⁻. Now what happens? Both K⁺ and A⁻ would like to equilibrate—that is, to move from an area of high concentration (inside the cell) to one of lower concentration (outside the cell). Potassium can pass across the membrane, but A⁻ cannot. Some K⁺ leaves the cell, and for every K⁺ that exits the cell, an extra negative charge is left inside and an extra positive charge is added to the outside of the cell. A voltage difference thus develops between inside and out; the inside of the cell is more negative (has excess negative charges) than outside the cell (which has excess positive charges).

Establishment of the resting potential across a cell membrane is somewhat more involved than the simple situation just described, but K⁺ is the principal ion that determines the resting potential of virtually all nerve cells, and the principles illustrated by this example are correct. If K⁺ levels

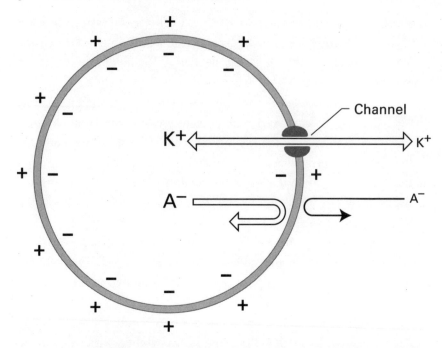

FIGURE 2.3 *A simplified model of a cell showing how a resting potential is established. The concentrations of K^+ and A^- are higher inside the cell than outside; thus, diffusion pressure to move from inside to outside the cell exists for both ions. Channels are present that let K^+ cross the membrane, but none that allow A^- to pass. Some K^+ ions exit the cell because of the diffusion pressure, causing the buildup of negative charge inside and positive charge outside. This difference in charge is the resting potential.*

on the outside of nerve cells rise and thereby reduce the difference in K^+ concentration between inside and out, the pressure on K^+ ions to move from inside to outside decreases. The result is a smaller resting voltage across cell membranes, resulting in the generation of abnormal electrical signals by nerve and muscle cells. This happens in kidney disease, as was described at the beginning of the chapter: K^+ levels in the blood rise, which means that nerve and muscle cells fail to maintain normal resting potentials and to generate proper electrical signals. The heart is especially affected, and unless the K^+ levels in the blood are lowered—for example, by treatment with an artificial kidney (a process called dialysis)—death will ensue because the heart cells no longer produce coordinated electrical signals.

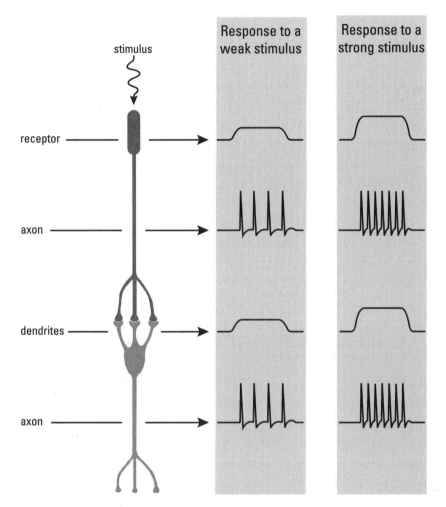

FIGURE 2.4 *Electrical potentials generated by neurons. A stimulus to a receptor (left) elicits receptor potentials (right) whose amplitudes are graded (larger or smaller) depending on stimulus strength. Transient action potentials, whose amplitude is constant but whose frequency alters in response to stimulus strength, are generated along the receptor's axon. Synapses between the receptor and adjacent neuron (left) results in the generation of graded synaptic potentials in the dendrites of the neuron (right). Information is transmitted down the neuron's axon via action potentials.*

Neuronal Signaling

Two types of electrical signals are generated by neurons: *receptor* and *synaptic potentials* and *action potentials*. Receptor and synaptic potentials are similar and are evoked when a sensory stimulus or a synaptic input impinges on a neuron. These potentials in turn generate action potentials, which carry information along axons. Figure 2.4 presents an idealized scheme illustrating these relations. Sensory stimuli produce potentials in receptor cells, and the receptor potentials generate action potentials along an axon. When an action potential reaches a synapse, a chemical is released and produces a synaptic potential in the adjacent neuron. This in turn generates an action potential in the axon of that neuron, and the information travels on to yet another cell. Eventually, neurons in the chain will impinge on effector cells, such as muscle cells, and a behavior will result.

Receptor and synaptic potentials have many common properties and can be considered together. Our model will be a synaptic potential that enhances activity in a contacted cell. Synapses producing such potentials are called *excitatory synapses*. (Synaptic potentials that depress the activity of a cell also occur widely in the brain; they are produced at *inhibitory synapses*.) At excitatory synapses, specific channels are present in the membrane (the *postsynaptic membrane*) across from where the chemical is released. In the absence of released chemical, the channels in the postsynaptic membrane are closed. But when the synapse is active, the chemical floods the space between the two cells, interacts with the channels, and causes them to open and allow mainly Na^+ to pass across the membrane. Since Na^+ has a positive charge, the inside of the postsynaptic cell becomes more positive as Na^+ flows in. Similar excitatory signals to a cell are seen in response to both sensory and synaptic input to cells.

The stronger the input, the larger the signal generated; that is, more ions flow into the cell and the voltage change across the membrane is greater (Figure 2.4, right). Thus, the strength of the input signal is coded in terms of the amount of voltage change; this is a graded signal. Changes in membrane voltage of 10–50 millivolts can occur at excitatory synapses and at sensory receptor sites.

The second type of electrical signal generated by neurons, the action potential, also results from Na⁺ flowing through channels in the membrane. Yet these channels open in response not to transmitter or a sensory stimulus but to a change in voltage across the membrane (Figure 2.5). At resting membrane voltage (when a neuron's membrane potential is negative 70 millivolts inside (indicated as -70 millivolts), the channels are closed, but when the membrane voltage becomes more positive (termed depolarization), the channels begin to open and admit Na⁺.

As the membrane voltage becomes more positive, more Na⁺ channels open, letting more Na⁺ into the cell (Figure 2.5a). As Na⁺ comes across the membrane, the inside of the cell becomes more positive still, which opens even more Na⁺ channels, and soon all the Na⁺ channels in that part of the membrane rapidly open. This is a positive-feedback (or regenerative) system that results in a substantial change in membrane voltage (about 100 millivolts) occurring in less than a millisecond (1/1,000 of a second). The voltage-sensitive channels, once opened, do not stay open

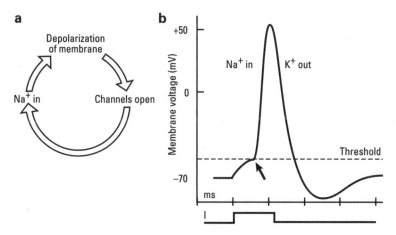

FIGURE 2.5 *(a) Generation of an action potential. Depolarization of the membrane causes Na⁺ channels to open, allowing Na⁺ into the axon, which causes more membrane depolarization, more Na⁺ channels opening, and more Na⁺ entering the axon. A positive feedback loop is established. (b) Depolarization of the axonal membrane by a current pulse (lower trace) opens enough Na⁺ channels to initiate an action potential. At threshold (dashed line), Na⁺ rushes in, causing the membrane potential to rapidly depolarize to about +50 millivolts arrow. The Na⁺ channels then spontaneously close after about 1 millisecond, and the membrane potential returns to −70 millivolts because K⁺ flows out of the cell.*

very long. Within another millisecond or so they close spontaneously, and recovery of membrane voltage to resting levels then occurs over the next 2 milliseconds or so, by K⁺ leaving the cell via channels that are also voltage-sensitive, but that open more slowly than the voltage-sensitive Na⁺ channels, and allow K⁺ to leave the cell rapidly.

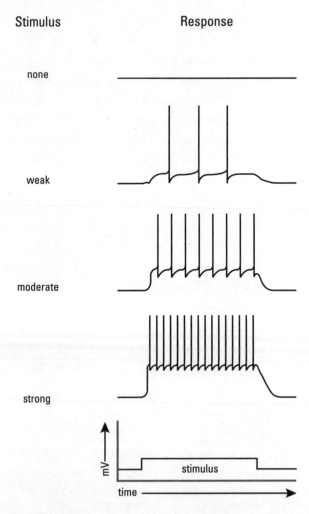

FIGURE 2.6 *The relation between receptor-synaptic potentials and action potentials. The greater the amplitude of the receptor-synaptic potential, the greater is the frequency of the action potentials generated. Weak stimuli generate small receptor-synaptic potentials, and correspondingly fewer action potentials are generated. Strong stimuli generate large receptor-synaptic potentials and many more action potentials.*

Action potentials are typically generated in a neuron close to where its axon arises and are all-or-nothing electrical events. To initiate an action potential requires a change of membrane voltage of about 15 millivolts, but once this threshold level is reached (arrow in Figure 2.5b), every action potential generated is exactly the same size. In neurons, the membrane voltage change that generates action potentials is provided by the synaptic and receptor potentials, hence the term *generator potential.*

If all action potentials are exactly the same size, how do they code for stimulus strength? Not by size of response but by frequency. The stronger the stimulus to a neuron, the more action potentials are generated per unit time. A weak stimulus generates few action potentials; a strong one many more action potentials per unit time. Figure 2.6 shows the relations between receptor-synaptic potentials and action potentials in the nervous system. In receptors and in dendrites, potentials are generated whose sizes are graded according to stimulus strength; along axons, action potential frequency codes stimulus strength. In receptors and in dendrites, therefore, information is coded by an amplitude modulation (AM) system, whereas axons code information by a frequency modulation (FM) system. Radio signals are also coded in these two ways—certain radio stations send AM signals, others FM signals—and with the flick of a switch most radios can receive either type of signal.

Transmission Down Axons

Why do axons generate action potentials—or, why do neurons generate two types of potentials? The answer is simple. Action potentials propagate themselves, but receptor and synaptic potentials do not. This means that once an action potential is generated, it moves along the axon without a change in amplitude. It remains the same size from the beginning to the end of the axon, even if the axon is several feet long! The axon potential is thus ideal to carry information long distances, which in the nervous system is a distance of more than 1–2 millimeters. Receptor and synaptic potentials, though, become smaller and smaller as they move away from their generation site.

How does the action potential propagate itself along an axon? This results simply from the way action potentials are generated; that is, an action potential is initiated when a change of about 15 millivolts occurs across a bit of the cell membrane. But once the action potential is initiated, it rapidly becomes full size, because more and more Na⁺ channels open. The Na⁺ passing through the channels quickly makes the axon more positive inside. Once inside, the Na⁺ moves along the axon, resulting in making the charge under the next bit of membrane more positive (Figure 2.7). As this happens, Na⁺ channels in that part of the membrane

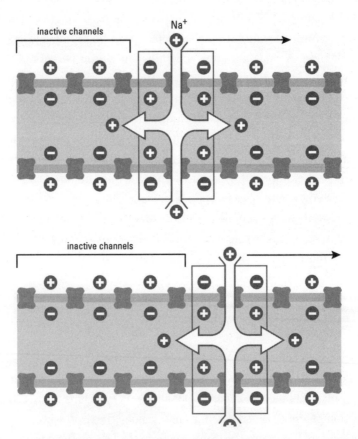

FIGURE 2.7 *The propagation of action potentials along an axon. When an action potential is generated (top, boxed area), positively charged Na⁺ ions flow into the axon. This results in positive charges moving down the axon, making the inside of the axon more positive. This results in the opening of Na⁺ channels in adjacent parts of the membrane and ultimately in the generation of another action potential (bottom, box area). The action potential moves only in one direction (to the right) because channels recently activated (to the left) are inactive for a short time—they will not open in response to changes in membrane voltage.*

begin to open, resulting in further voltage change, and soon threshold is reached. Hence, action potentials are generated all along the length of an axon once a spike is initiated in the axon.

An interesting question is why spikes travel only in one direction along axons. The reason is that once a bit of membrane has generated an axon potential, it cannot be excited again for a few milliseconds. In other words, the channels do not open in response to voltage for a short time. This is a long enough time that the action potential moving down the axon is sufficiently far away not to affect significantly that part of the membrane whose channels are once again active.

Action potentials travel down axons at rates of 100–200 miles an hour in animals such as ourselves. In *invertebrates* (animals without backbones), action potentials travel at much slower speeds, 30–40 miles an hour. Glial cells contribute to the rapid rate of action potential travel down vertebrate axons by forming an insulating layer, called *myelin*, around axons (Figure 2.8). They do this by wrapping many layers of membrane around the axons. The myelin membrane has little protein in it, hence few channels. Thus, ions cannot easily cross myelin; it acts as a good insulator. Myelin not only speeds action potential conduction but also makes action potential propagation more efficient by limiting the generation of new action potentials to certain regions—*nodes*—along the axon where there are gaps in the myelin. Thus, action potentials are not generated continuously along vertebrate axons; rather, they are generated only at the nodes. So action potential generation jumps from node to node, and this works as long as the nodes are not too far apart. The advantage is that much less energy is needed to transmit action potentials down vertebrate axons compared with invertebrate axons (which do not have myelin), and action potentials travel much faster down vertebrate axons.

If myelin is damaged or diseased, propagation down axons is compromised, which happens in *multiple sclerosis (MS)* and several other demyelinating conditions. MS is one of the most common central nervous system diseases of young adults, affecting about one in a thousand individuals. It strikes mainly between the ages of 25 and 40, and about two-thirds of the cases occur in women. Early symptoms include blurred vision, uncoordinated walking, numbness, and fatigue. In later stages, patients may exhibit slurred speech, tremors, memory loss, and severe paralysis.

The cause of MS is not known, but it is thought to be an autoimmune disease. For unknown reasons, the immune system of those affected forms antibodies against myelin, causing it to disintegrate. The disease is also capricious—patients typically have remissions and relapses that can occur over a period of years.

Why does MS cause neurological symptoms? When myelin is lost, action potential conduction is first slowed and eventually may be entirely blocked. In essence, the loss of myelin has the same effect as increasing the distance between the nodes, compromising action poten-

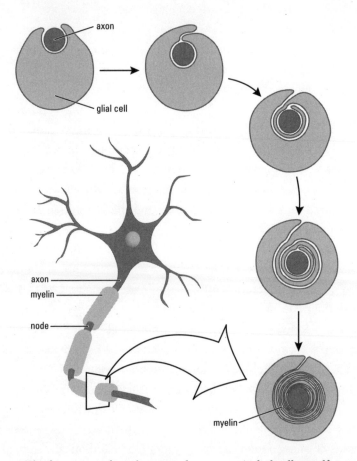

FIGURE 2.8 *The formation of myelin around an axon. A glial cell engulfs an axon and then wraps many layers of membrane around the axon. The cytoplasm between the layers of membrane gets squeezed out, leaving a highly compacted myelin sheath. The myelin sheath is not continuous along an axon but is interrupted by gaps or notes, as shown in the drawing at the bottom left.*

tial generation. If enough myelin is lost, action potentials no longer can jump the gap between nodes. A slowing or even a partial block of action potential transmission down axons has a devastating effect on neural activity and behavior.

Synapses

I have so far dwelled on excitatory synapses. At such synapses, channels allow mainly Na^+ to flow into the cell, which causes the inside of the cell to become more positive and the cell membrane to approach action potential threshold. For this reason these synapses are called excitatory; they cause the cell to generate action potentials.

But, as noted earlier, there are inhibitory synapses as well. These synapses make a neuron less likely to fire an action potential. How do they work? In a fashion similar to excitatory synapses, but at inhibitory synapses a different ion flows across the membrane. Channels at inhibitory synapses typically allow Cl^- to pass into the cell; since Cl^- is negatively charged, the cell becomes more negative inside when inhibitory synapses are activated. This causes membrane voltage to move away from the action potential threshold. Inhibitory synapses thus slow down or prevent action potential generation.

The rate of action potential generation by a neuron depends on the interplay of excitatory and inhibitory synapses onto the cell, as shown schematically in Figure 2.9. In Figure 2.9a, excitatory synapses impinge on the cell's dendrites; inhibitory synapses are on the cell body. Action potentials are first generated by the membrane adjacent to the axon. Surrounding the axon is myelin, interrupted by nodes that generate the action potentials traveling down the axon.

Figure 2.9b is a record of membrane voltage, as would be recorded by placing an intracellular recording probe into the cell and making measurements over time. When the probe enters the cell, a resting potential of about −70 millivolts is recorded. When excitatory input to the neuron is applied (depicted as ticks along the line marked "excitatory"), small excitatory synaptic potentials are elicited in the dendrites. These add together, and with repetitive excitatory stimuli they cause a buildup of

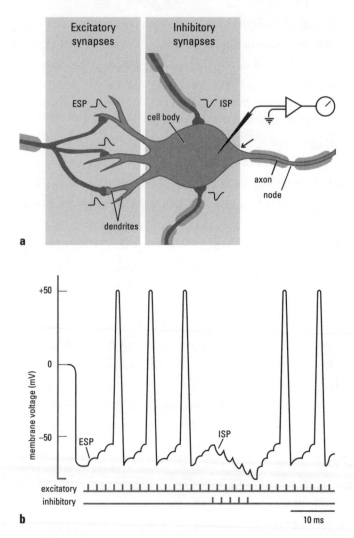

FIGURE 2.9 *How a neuron is excited and inhibited by synaptic input. (a) Excitatory synapses are found predominantly on dendrites; inhibitory synapses, on the cell body. Each active synapse generates a small positive excitatory synaptic potential (ESP) in the dendritic tree or a small negative inhibitory synaptic potential (ISP) in the cell body. (b) A recording made from the cell body will record the resting, synaptic, and action potentials. The resting potential of the cell is approximately –70 millivolts. Excitatory input elicits excitatory synaptic potentials (ESPs), which sum, causing the membrane potential to become more positive. This results in the generation of action potentials where the axon arises (arrow). Activating the inhibitory input elicits inhibitory synaptic potentials (ISPs), which cause the membrane potential to become more negative. As this happens the neuron ceases firing. Once the inhibitory input stops, the cell membrane becomes more positive and again fires action potentials.*

positive charge within the cell that brings membrane potential to threshold for action potential generation.

As long as excitatory input continues, the cell will generate action potentials, one after another. But if inhibitory input to the cell is activated (ticks along the line marked "inhibitory"), the membrane potential becomes more negative (inhibitory synaptic potential). The membrane potential falls below action potential firing threshold levels and the cell stops firing. Once the inhibitory input stops, the excitatory input drives the membrane potential to the action potential threshold again, and the cell generates action potentials as long as excitatory input continues. Thus, the rate of spike firing in a neuron reflects the balance of excitatory and inhibitory input to a neuron, and the message traveling down the axon reflects this balance.

Synaptic Mechanisms

Among the key structures at a synapse are the *synaptic vesicles* that store the chemicals to be released, which at excitatory and inhibitory synapses are called *neurotransmitters*. When a synapse becomes active, the vesicles join to the membrane and release the stored neurotransmitter. How does this happen?

When an action potential travels down an axon and reaches a *synaptic terminal*, the inside of the terminal becomes more positive because of the Na^+ flowing across the membrane through Na^+ channels. In the terminal membrane are other channels that respond to this voltage change; they open and admit an ion that has two extra positive charges, *calcium* (Ca^{2+}). Calcium ions promote the docking of vesicles to the membrane. This results in the fusion of the vesicle to the membrane and the opening of the vesicle to the outside. Neurotransmitter is then released and flows to channels on the postsynaptic membrane, thus opening them.

Figure 2.10 shows two other important features of synaptic transmission. First, after the synaptic vesicles bind to the presynaptic membrane and release their contents, they become incorporated into the terminal membrane. But away from the active zone of the terminal (where neurotransmitter is released), new vesicles are formed by the inpocketing of membrane. Thus, vesicles recycle in the synaptic terminal.

The other important aspect of synaptic transmission illustrated in Figure 2.10 is the breakdown or reuptake of neurotransmitter. Once neurotransmitter is released from the vesicles, it is necessary to get rid of it

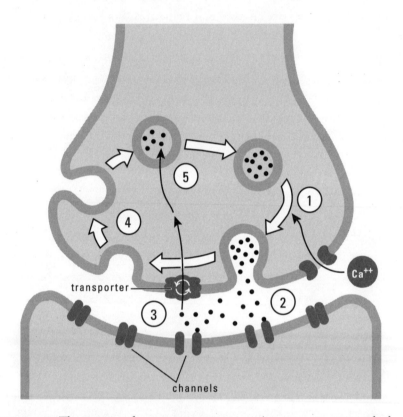

FIGURE 2.10 *The process of synaptic transmission. A synapse is activated when the terminal membrane becomes more positive as a result of an action potential reaching the terminal. This causes voltage-sensitive Ca^{2+} channels to open, admitting Ca^{2+} into the terminal (1). Ca^{2+} facilitates the binding of synaptic vesicles to the membrane. The vesicles release their neurotransmitter content, which diffuses to the postsynaptic membrane and activates channels there (2). Neurotransmitter is either broken down by enzymes (not shown) or taken back up into the terminal by transporters (3), allowing its reuse. After fusing with the membrane, the synaptic vesicles become incorporated into the membrane (4). New synaptic vesicles form by an inpocketing of the terminal membrane. The newly formed synaptic vesicles are filled with neurotransmitter (5) and the process is ready to go again.*

rapidly so newly released neurotransmitter can have effects. This occurs in two ways. Some neurotransmitters are broken down by enzymes to inactive products; most neurotransmitters, though, are rapidly taken back into the terminal by membrane proteins called *transporters*. These transporters move the neurotransmitter into the terminal, where it is repackaged into the newly formed vesicles. There is much interest in transporters among neuroscientists because potent drugs such as cocaine and Prozac inhibit transporter function. Inhibition of a transporter allows a neurotransmitter to remain in the synaptic cleft for a longer period of time. In essence, it is as though one has increased the amount of neurotransmitter released at the synapse, and there is now abundant evidence that altering the levels of neuroactive substances at synapses can cause specific psychological changes.

Vulnerability of Synapses

Synapses are vulnerable to blockage, and many drugs and poisons that affect the nervous system disrupt one or another aspect of their ability to transmit or receive signals. Indeed, drugs affect virtually every aspect of synaptic transmission. For example, a toxin from the deadly bacterium *Clostridium botulinum* that causes botulism potently blocks synaptic transmission by preventing the release of neurotransmitter from synaptic terminals. The toxin does this by interfering with the binding of the synaptic vesicles to the membrane of the terminal. Proteins present in both the synaptic vesicle and cell membranes are involved in the attachment and fusion of the vesicles to the cell membrane that results in the release of neurotransmitter at a synapse. The botulinum toxin acts by breaking down at least one of these attachment proteins. Thus, the vesicles cannot fuse to the membrane and release their neurotransmitters.

Clostridium botulinum is found commonly in soil and on fruits and vegetables. It is an anaerobic bacterium—that is, it thrives and multiplies under conditions of low oxygen. It can become a serious problem to humans who eat canned fruits and vegetables that have not been heated enough during the canning process to destroy the bacterium. If any bacteria remain in the low-oxygen atmosphere of a can or jar, they multiply and produce copious amounts of the toxin. Eating the tainted food can cause death by blocking synaptic transmission throughout the nervous system; a person so poisoned cannot breathe, for example. Botulinum toxin poisoning was much more of a problem in the early days of canning, but even today occasional incidents occur. Botulinum toxin can be made fairly easily and is exceptionally potent; the worry is that it may be used as a biological warfare agent. Indeed, it was one of the weapons of mass destruction Saddam Hussein was thought to have that led to the Iraq invasion by the United States in 2003. One estimate predicts that as many as 40,000 people could be killed by just 200 pounds of botulinum toxin. Tetanus toxin is related to the botulinum toxin and also acts by preventing the release of neurotransmitters at synapses.

Synaptic transmission can also be blocked by preventing a neurotransmitter from interacting with the channels in the postsynaptic membrane. *Curare* is one example of substances that have this effect. It is a paralyzing agent that blocks synaptic transmission between nerve and muscle at the neuromuscular junction, rapidly causing paralysis and death. This naturally occurring substance, found in certain climbing vines, was discovered by South American Indians, who used it to coat the tips of arrows and spears. Another agent that blocks the neuromuscular junction by preventing neurotransmitter from interacting with membrane channels is α-bungarotoxin, a deadly component of cobra snake venom. It binds tightly to the channels, essentially destroying them. Once α-bungarotoxin attaches to a channel, transmitter can no longer interact with the channel and activate it. Like curare, α-bungarotoxin causes muscle paralysis and death.

Agents can also prevent the breakdown of neurotransmitter after its release or prevent reuptake of the transmitter into the terminal. Examples of the former are certain organic phosphates, which are the principal ingredients in many pesticides and in deadly nerve gases such as sarin, the gas that killed a dozen people in the Tokyo subway station terrorist attack in 1995. Very recently, sarin gas was believed to be used by Syrian President Assad's military in that country's civil war. A tiny amount of these agents causes very rapid paralysis by blocking the enzyme that breaks down the neurotransmitter at the neuromuscular junction. Neurotransmitter builds up excessively at the synapse, and the synapse can no longer function properly. Another drug that acts the same way is eserine, a derivative of the African calabar bean. Eserine was used in some parts of Africa as a truth serum. The legend is that a person who was unjustly accused would rapidly drink down the mixture. Since eserine also causes vomiting, such an individual would vomit much of the poison and survive. The guilty individual would drink the truth serum more slowly, would absorb more of the poison, and would die.

Some substances prevent or inhibit the reuptake of neurotransmitters. As noted above, neuroactive drugs such as cocaine and Prozac exert their effects by inhibiting the transporters involved in reuptake of neurotransmitters. If reuptake is entirely prevented, synaptic transmission will soon be completely shut down as neurotransmitter floods the synaptic cleft, and dire consequences will result. Neuroactive drugs such as cocaine and Prozac inhibit reuptake processes only partially—they simply raise neurotransmitter levels in the synaptic cleft. Furthermore, they affect only the uptake of specific substances released at synapses.

A Synaptic Disease: Myasthenia Gravis

Myasthenia gravis, a disease of the synapses between nerve and muscle, is characterized in its early stages by muscle weakness and fatigue. Diplopia (double vision), due to a weakness of the ocular muscles, is another early symptom, along with droopy eyelids. In its latter stage, patients may be bedridden and even die. Fortunately, myasthenia gravis is a relatively rare disease; it affects about 25,000 people in the United States. The diagnosis, treatment, and analysis of this disease have depended on a detailed understanding of what happens during synaptic transmission at the neuromuscular junction and the drugs that block transmission at this synapse.

Many years ago, a patient suffering from myasthenia gravis was given a tiny amount of curare. The drug induced muscle weakness and fatigue far beyond that expected based on observations of normal people given similar amounts of the drug. These findings provided some of the first evidence that myasthenia gravis is a disease involving the synapses between nerve and muscle. The excessive sensitivity of myasthenia gravis patients to drugs such as curare has been used to diagnose the disease. Yet giving a patient eserine, which prevents breakdown of the neurotransmitter at the neuromuscular junction, is used to help patients with the disease. Their symptoms are alleviated, and indeed, eserine and eserine-like drugs are the primary therapy for the disease.

Why would curare accentuate and eserine relieve symptoms of the disease? Experiments in the mid-1960s showed that patients with myasthenia gravis have fewer excitatory channels in the postsynaptic membrane at muscle synapses than do normal individuals. Curare, by interfering with the interaction of neurotransmitter with channels, exaggerates the deficit by reducing the number of activated channels. Eserine, by contrast, interferes with the breakdown of neurotransmitter and allows it to remain intact longer at the synapse. Thus, in the presence of eserine, channels are activated longer than normal. This excites the muscle more, as though more channels were present in the postsynaptic membrane.

Finally, why are there fewer neurotransmitter-activated channels in patients with myasthenia gravis? There is now good evidence that myasthenia gravis, like multiple sclerosis, is an autoimmune disease. Individuals with myasthenia gravis form antibodies against the channels in the postsynaptic muscle membrane. Why this happens is not understood, but what is known is that the antibodies attach tightly to the channels, leading to their destruction, much as α-bungarotoxin does. Ordinarily, neurotransmitter-activated channels are synthesized continuously by muscle (and nerve) cells, replacing old ones or ones that have been damaged for one reason or another. Indeed, neurotransmitter-activated channels at the neuromuscular junction are completely replaced every week or so. In myasthenia gravis, the antibodies inactivate the channels as fast as or faster than they can be made, hence the lower number of functional channels at the neuromuscular junction in these patients.

An understanding of the causes of myasthenia gravis suggests a strategy for its cure. If antibody levels in the blood can be reduced, the number of functional channels in the postsynaptic membrane should increase. It is possible to reduce antibodies in a patient's blood by treatment with the artificial kidney (dialysis), and spectacular improvement is seen in such patients. Unfortunately, the effect is only temporary; within a few days after dialysis, antibody levels increase. At present there is no way to lower antibody levels permanently, but if this active area of research can be made successful, patients with myasthenia gravis and similar autoimmune diseases could be completely cured.

3

Neuromodulation, Drugs,
and the Brain

Tess was the eldest of ten children born to a passive mother and an alcoholic father. She was abused in childhood in both the physical and sexual senses. When Tess was twelve, her father died, and her mother entered a clinical depression from which she never recovered. Tess—one of those inexplicably resilient children who flourish without any apparent source of sustenance—took over the family.

At seventeen, she married an older man, in part to provide a base for her brothers and sisters. The husband became alcoholic, and was abusive when drunk. The marriage became loveless and collapsed once the children were grown.

Meanwhile, Tess had made a business career out of her skills of inspiring others and achieved a reputation as an administrator capable of turning around struggling companies. She rose to a high level in a large corporation. . . .

Her personal life was unhappy, and she stumbled from affairs with abusive married men. When these ended, she was severely demoralized. A current affair had lasted months, and she was now less energetic and more unhappy. When I first met her, she surprised me, utterly charming and a pleasure to be with. However she had all the signs and symptoms of depression: tears, sadness, absence of hope, inability to experience pleasure, loss of sleep and appetite, guilty ruminations. . . .

Two weeks after starting Prozac, Tess was no longer feeling weary. She had almost not known what it was to feel rested and hopeful. She had been depressed much of life and was astonished at being free of depression. She looked different, more relaxed and energetic. She laughed more frequently, and its quality was different, lively, even teasing.

With this new demeanor came a new social life, one that appeared instantly and full blown.

—*Excerpted from Peter D. Kramer, Listening to Prozac (New York, NY: Penguin, 1997)*

At synapses, neurons can not only be excited or inhibited but also be modified or modulated in a variety of ways. This second type of synaptic action is called neuromodulation. I have already described how neurons are excited or inhibited at synapses; now I'll detail how neurons are modulated by synaptic action. Why are these neuromodulatory processes important? Because long-term changes that occur in the brain and result in phenomena such as remembering and learning are thought to happen because of neuromodulatory synaptic action. But also, many chemicals that alter mental states, such as LSD and amphetamines, and diseases that cause thought disorders, such as schizophrenia, can be related to alterations in neuromodulatory synaptic transmission. And as the story of Tess illustrates, inducing alterations in levels of neuromodulatory substances in the brain by drugs such as Prozac can provide striking relief from serious psychiatric disorders such as deep depression.

We know that transmission at most brain synapses is chemical. A substance released from one neuron diffuses to an adjacent neuron and induces changes in that neuron. A few synapses are electrical; at such synapses the cell bodies of two adjacent neurons become closely apposed, and protein channels create a bridge between the contacting cells. These channels enable ions to flow directly from one cell to another. A change in membrane voltage in one cell results in a rapid change in voltage in the contacted cell as the ions flow from one cell to the next. Such transmission is very fast and usually can go in either direction. *Electrical synapses*, then, synchronize activity between neurons or permit reciprocal interactions between neurons.

But signals cannot be amplified at electrical synapses—amplification is possible only at chemical synapses. Also, signal polarity cannot be changed at electrical synapses—from excitation in one cell, for example, to inhibition in the adjacent cell or vice versa. Electrical synapses can be modified in their efficacy by neuromodulation, but chemical synapses are predominant in the brain because they allow for a richness of interactions between neurons.

Neurotransmitters and Neuromodulators

As many as fifty different substances may be released at chemical synapses in the brain. Usually no more than one or two substances are released at any one synapse, and it is generally believed that all the synapses made by a neuron release the same substance or substances. So neurons have a lot of diversity and specificity in relation to the substances they release at their synapses.

It is convenient to classify the substances released at synaptic sites into two categories, neurotransmitters and neuromodulators, based on their mode of action on postsynaptic cells. *Neurotransmitters* are substances that interact directly with channels in the postsynaptic membrane; they mediate fast excitatory or inhibitory synaptic action. When a neurotransmitter activates such channels, they open and allow certain ions to cross the membrane. As described in Chapter 2, when positively charged ions such as Na^+ cross the membrane, the inside of the cell becomes more positive and the neuron is more likely to generate an action potential—it is excited. Conversely, if negatively charged ions such as Cl^- cross the membrane, the inside of the cell becomes more negative and the cell is less likely to generate an action potential—it is inhibited. These excitatory and inhibitory interactions are quite fast. Once a synapse is activated, it takes less than half a millisecond (0.0005 second) for a voltage change to occur in the postsynaptic cell. Furthermore, the change in voltage in the postsynaptic cell lasts only a short time—between 10 and 100 milliseconds, or no more than 0.1 second.

Neuromodulators act quite differently on neurons. When neuromodulators are released at synapses, they diffuse to the postsynaptic membrane where they interact with membrane proteins called *receptors*, which are linked to intracellular enzyme systems. Activating these receptors does not directly open channels in the membrane; rather, the neuromodulators' action is mediated by biochemical changes in the postsynaptic neuron. In a typical situation, intracellular enzymes are activated and synthesize small second-messenger molecules (the first messenger is the neuromodulator). Many aspects of neural cell function can then be altered, from the properties of the channel proteins in the membrane to the expression of genes in the nucleus. In this manner, profound physiological and structural changes can happen in a neuron as a result of neuromodulation.

Neuromodulatory effects typically have a slow onset, usually seconds, and then can last for long periods of time—minutes, hours, days, or even longer. As noted in Chapter 11, the evidence is accumulating that long-term changes in the brain, which underlie phenomena such as memory and learning, result from neuromodulatory synaptic interactions.

Some substances released at synapses act exclusively as neurotransmitters; others are solely neuromodulators. Most substances released at synapses can act as both: at some synapses they interact with membrane channels postsynaptically, while at other synapses they interact with membrane receptors linked to intracellular enzyme systems. *Acetylcholine* is a typical example. It is the neurotransmitter used at all synapses between nerve and muscle in vertebrates (the neuromuscular junction), but it can also activate neuromodulatory membrane receptors in the brain linked to biochemical systems. These two kinds of actions can be distinguished by pharmacological means; certain drugs will block one action but not the other, and vice versa. Curare blocks the action of acetylcholine at the neuromuscular junction, whereas the drug atropine blocks the effect of acetylcholine on neuromodulatory membrane receptors. Atropine has no effect on the neuromuscular junction, and curare does not block the neuromodulatory action of acetylcholine on membrane receptors. Some agents can specifically activate one or the other action of acetylcholine. Nicotine activates the membrane channels specific to acetylcholine, while muscarine activates acetylcholine receptors linked to enzyme systems. These two effects are often referred to as the nicotinic and muscarinic actions of acetylcholine.

Figure 3.1 shows how a neuromodulatory system works. The system chosen for illustration uses *adenylate cyclase*, an enzyme that synthesizes a second-messenger molecule called cAMP. A neuromodulator (the first messenger), released from vesicles in the presynaptic terminal, interacts with receptors on the postsynaptic membrane and thereby activates them. The activated receptors now interact with an intermediate protein, called a *G-protein*, which in turn interacts with adenylate cyclase and activates it.

Virtually all known membrane receptors linked to intracellular enzyme systems—including the light-sensitive molecules in photoreceptors cells, the odorant-sensitive proteins in olfactory cells, and the neuromodulatory receptors on neurons—interact first with G-proteins. A number of G-proteins are known that can activate (or inhibit) a variety of intracellular enzymes.

Regarding the system shown in Figure 3.1, activated adenylate cyclase promotes the conversion of the molecule *adenosine triphosphate (ATP)*, which is ubiquitous in cells and stores and provides energy, into a smaller molecule called *cyclic adenosine monophosphate (cAMP)*. cAMP exerts its effects by activating another kind of enzyme, termed a *kinase*, that adds a phosphate group to cellular constituents, usually proteins. This process, called *phosphorylation*, is a favorite way for cells to activate or inactivate biochemical reactions or to modify the properties of proteins.

The kinase activated by cAMP is called protein kinase A (PKA). It can cause effects at many levels of the cell, including the nucleus, the cytoplasm, and the membrane. In the nucleus, genes can be turned on or off; in the cytoplasm, enzymes can be activated or inactivated, including those involved in protein synthesis; and at the membrane, ion channels and other membrane proteins can be altered. For example, phosphorylation of a channel can modify its sensitivity to a neurotransmitter, how long the channel remains open after its activation by a neurotransmitter, or even its ionic specificity—that is, which ions it will admit into or out of the cell.

Another interesting feature of neuromodulation is that a substance released at a synapse can affect the terminal releasing it; thus, synaptic terminals can have their own receptors, termed *autoreceptors*, that respond to the neuromodulator. In our example in Figure 3.1, autoreceptors are linked to the cAMP cascade. The activated kinase may, for example, phosphorylate enzymes that synthesize the substance released

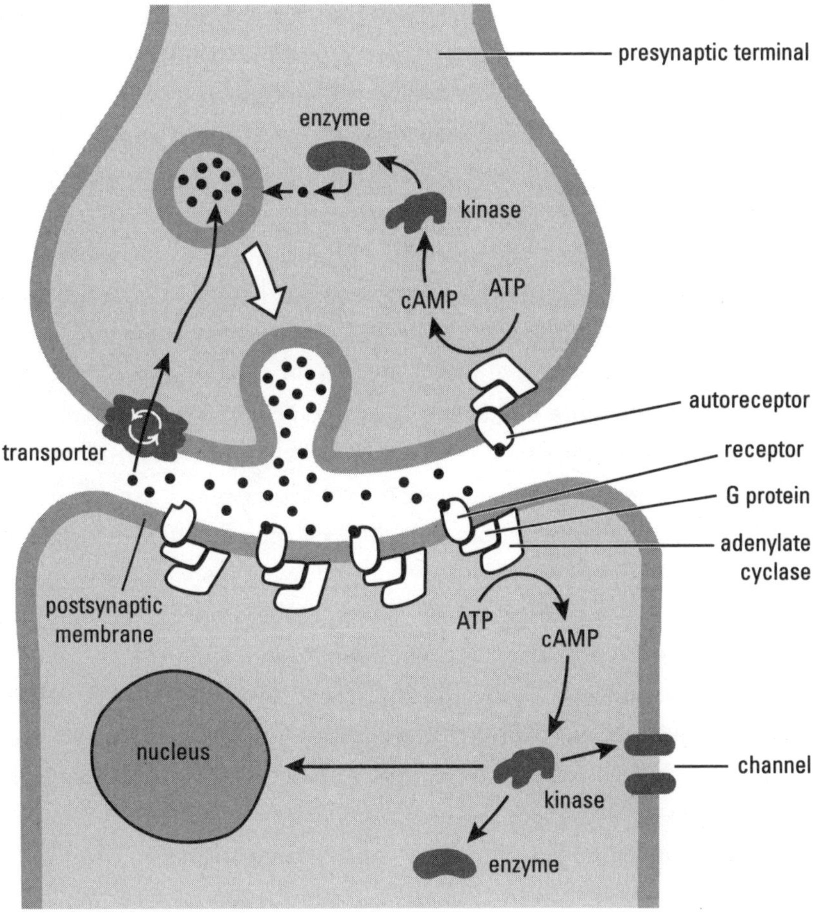

FIGURE 3.1 Neuromodulatory synaptic transmission. The activation of the presynaptic terminal, binding of vesicles to the terminal membrane, release of neuroactive substances from the synaptic vesicles, and reuptake of the released substances back into the terminal by transporters are identical to the processes shown in the scheme of synaptic transmission in Figure 2.10. However, at neuromodulatory synapses, the released substances bind to protein receptors on the postsynaptic membrane linked to a G-protein. In the case illustrated, the G-protein activates an enzyme, adenylate cyclase, that converts adenosine triphosphate (ATP) to cyclic adenosine monophosphate (cAMP). cAMP in turn activates a kinase that can act on channels in the membrane, on enzymes in the cytoplasm, or on proteins regulating gene expression in the nucleus. In addition, autoreceptors, found on the presynaptic terminal, may be activated. In the case shown, the cAMP cascade leads to the activation of an enzyme involved in synthesizing the neuroactive substance released from the synaptic vesicles of the terminal.

by the terminal. With such mechanisms a neuromodulator can regulate how much substance is made by the terminal; that is, a terminal can modify its own properties by feedback mechanisms.

Even though several second-messenger cascades are known, the list is probably far from complete. cAMP is the most extensively studied and best-known *second messenger*. Other second messengers activate other kinases, and kinases are very specific with regard to the molecules they can phosphorylate. One neuromodulator at a synapse can activate more than one second-messenger pathway by way of one receptor but several G-proteins; thus, a variety of processes can be altered in a cell due to a single neuromodulatory input. Any neuron probably receives many such inputs, so the possibilities for its mode of modulation are considerable.

The distinction between synaptic neurotransmission and neuromodulation is sometimes blurred. Some ion channels in the membrane activated by neurotransmitters can admit Ca^{2+} ions into neurons, and the admitted Ca^{2+} can act as a second messenger. The way this works is that the Ca^{2+} binds to a protein within the cell, termed *calmodulin*, and the Ca^{2+}-calmodulin complex interacts with specific kinases, termed CaM-kinases . These kinases act like other second-messenger-activated kinases, phosphorylating cellular constituents and thus activating or inactivating biochemical mechanisms. Long-term changes in vertebrate neurons that relate to memory and learning appear to be mediated by such a mechanism (see Chapter 11).

Classifying Synaptic Substances

Although as many as fifty substances may be released at synapses in the human brain, they fall into one of four chemical classes. Two of these classes act mainly as neurotransmitters (acetylcholine and amino acids), and the other two act mainly as neuromodulators (*monoamines* and *neuropeptides*). I'll start with the neurotransmitters.

Acetylcholine is the only naturally occurring substance in its class, but other chemicals mimic acetylcholine activity; they will activate acetylcholine channels or receptors. Acetylcholine is usually an excitatory neurotransmitter, but it can exert inhibitory effects through neuromodulatory actions. Indeed, its discovery in the early 1920s by Otto Loewi, a German neurochemist, came about in relation to a slowed heart rate. Acetylcholine is released around the heart when the vagus nerve, which regulates the heart, is active; acetylcholine interacts with a receptor on heart cells linked to a G-protein. The G-protein inhibits adenylate cyclase and thus the synthesis of cAMP. The fall in cAMP levels diminishes the phosphorylation of certain channels in heart cell membranes, thereby slowing down the heart (see Figure 12.5). Loewi stimulated the vagus nerve in an animal and collected the fluid around its heart. He demonstrated that this fluid slowed the heart of another animal whose vagus nerve was not stimulated. It was learned later that the substance released by the vagus nerve is acetylcholine.

Like acetylcholine, the *amino acids* released at synapses act mainly as neurotransmitters, although some can interact with membrane receptors linked to second-messenger cascades. One amino acid, *glutamate*, serves as the major excitatory neurotransmitter in the brain; two other amino acids, *glycine* and *γ-aminobutyric acid (GABA)*, are the major inhibitory neurotransmitters in the brain.

Glutamate, GABA, and glycine are closely related structurally, as you can see in Figure 3.2. The molecule glutamate becomes GABA when a carbon atom and two oxygen atoms are removed. Without its tail of $-CH_2-CH_2-COOH$, the molecule is glycine.

Glutamate activates two types of channels. One allows Na^+ to enter a cell, thus exciting it. The other channel allows both Na^+ and Ca^{2+} to enter the cell, which excites the neuron and also activates calmodulin and CaM-kinases. Thus, when a neuron possesses this channel, it can be both excited and modulated at the same synapse. As noted earlier, there is evidence that this channel plays an important role in storing memories. When we experience something, not only are our neurons excited, but they must also be modified in some way to account for memory. How this happens is not completely understood, but neuromodulatory pathways activated at these glutamate synapses appear to underlie such long-term changes in neurons (see Chapter 11).

GABA and glycine mainly activate channels that allow Cl^- to flow into neurons. This results in inhibition of the neurons. The GABA-activated Cl^- channel is of particular interest because its properties are specifically altered by three kinds of drugs: barbiturates, benzodiazepines, and alcohol. Barbiturates and benzodiazepines are used to treat anxiety, and at low concentrations alcohol can also relieve anxiety and promote relaxation. The benzodiazepines were introduced in the 1960s, and by the early 1980s they were used by enormous numbers of people. A variety of benzodiazepines are available; the best known is diazepam (Valium). By the early 1980s, about 20 percent of the women and 10 percent of the men in England were taking a benzodiazepine at one time or another during the year.

Barbiturates, benzodiazepines, and alcohol all potentiate the GABA-activated Cl⁻ channels. When these drugs are present, the channels allow more Cl⁻ to cross the membrane; hence, the postsynaptic neurons are more strongly inhibited. How increased inhibition relieves anxiety is not understood. What is known is that barbiturates and benzodiazepines interact at specific sites on the GABA channel, and the sites are different from the site where GABA interacts with the channel protein. The fact that there exist sites specific for a drug on a membrane channel or receptor suggests that there is an endogenous brain substance that interacts with the site. As yet, no endogenous benzodiazepine or barbiturate-like substance has been identified.

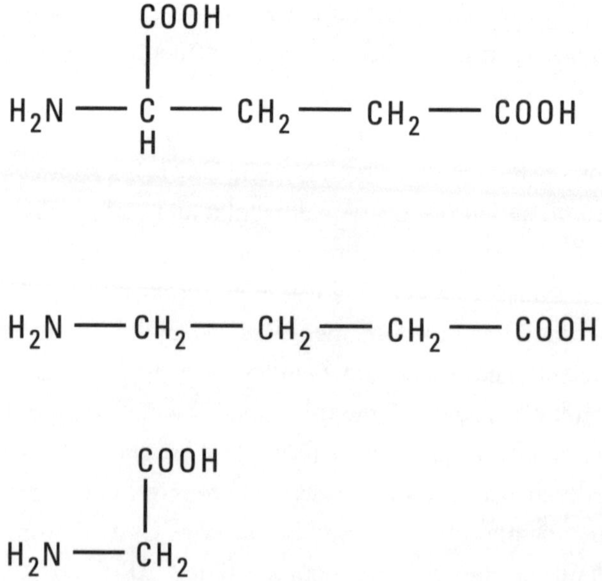

$$H_2N - \underset{\underset{H}{|}}{\overset{\overset{COOH}{|}}{C}} - CH_2 - CH_2 - COOH \qquad \text{glutamate}$$

$$H_2N - CH_2 - CH_2 - CH_2 - COOH \qquad \text{GABA}$$

$$H_2N - \underset{}{\overset{\overset{COOH}{|}}{CH_2}} \qquad \text{glycine}$$

FIGURE 3.2 *The molecular structures of the three most common amino acid neurotransmitters used in the brain. Glutamate, an excitatory neurotransmitter, is very similar in structure to GABA, an inhibitory neurotransmitter.*

Monoamines (of which there are two types—*catecholamines* and *indoleamines*) serve almost exclusively as neuromodulators. Catecholamines are derived from the amino acid *tyrosine*, and three catecholamines—*dopamine, norepinephrine*, and *epinephrine*—are important in brain function. Indoleamines are derived from the amino acid *tryptophan*, and one indoleamine, *serotonin*, is a key substance released at brain synapses. Figure 3.3 presents the structure of dopamine and serotonin and the amino acids from which they derive.

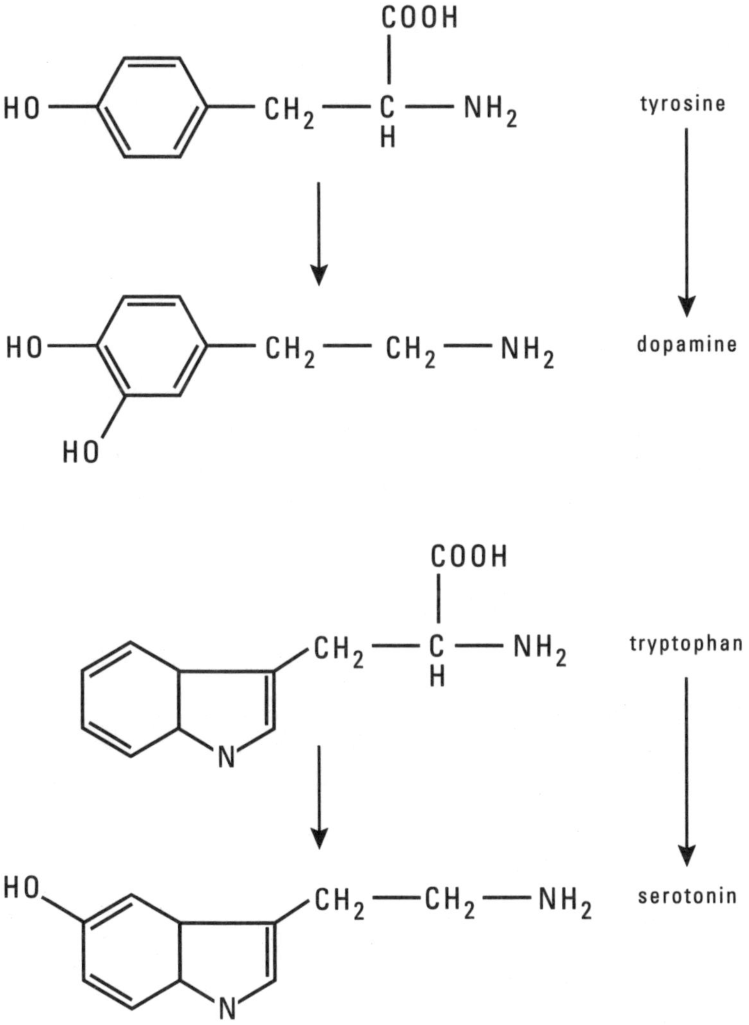

FIGURE 3.3 *Dopamine (top) and serotonin (bottom), two neuromodulatory substances, are derived from the amino acids tyrosine and tryptophan.*

Monoamines are critical in regulating the brain's mood, that is, its affective and arousal states. Drugs that alter transmission at synapses using one of the monoamines, or that alter the levels of these substances at synaptic sites, often dramatically change a person's mood or other mental state. The hallucinogen LSD (lysergic acid diethylamide) is a good example, because it interferes with membrane receptors activated by serotonin, thereby preventing serotonin from interacting with the monoamine receptors. The amphetamines, another example, are brain stimulants that induce hyperactivity and inability to sleep. They increase levels of catecholamines at synapses, particularly dopamine, by releasing these substances from nerve terminals and by interfering with the reuptake of catecholamines into the presynaptic terminals.

Dopamine, Parkinson's Disease, and Schizophrenia

The monoamines' range of activities is broad and by no means entirely known. For example, two quite different diseases of the brain, *Parkinson's disease* and *schizophrenia*, have been related to altered dopamine levels. Parkinson's disease is primarily a disease of the motor system. Patients suffering from Parkinson's disease develop a severe tremor and have trouble initiating movements. They also become rigid, and their movements are characteristically slow.

In the late 1950s, it was discovered that the brain's dopamine content was low in patients suffering from Parkinson's disease. Much of the brain's dopamine is in nuclei, termed the *basal ganglia*, that are involved in the initiation of movement. In one basal ganglia nucleus, more than 90 percent of the dopamine may be lost in the disease.

In the early 1960s, a drug therapy for Parkinson's disease was introduced that raises the brain's dopamine levels. The drug, *L-dopa*, a precursor of dopamine, is effective because it passes readily from the blood to the brain. Dopamine itself does not do that; it is blocked by the blood-brain barrier, which excludes many substances (good *and* bad) from entering the brain. Once in the brain, L-dopa is converted to dopamine and increases the brain's levels of the drug.

In recent years, another therapy for Parkinson's disease has been tried, with mixed results. With this therapy, cells are transplanted into the brain to release dopamine. Such transplants are expected to raise dopamine levels. Though neural and nonneural cells have been tried, the best results so far have been with fetal cells derived from basal ganglia nuclei or stem cells induced to become dopamine-producing cells. Although the results are promising, cell transplantation is not yet as effective as drug therapy. In patients with severe Parkinson's disease who do not respond to L-dopa therapy, destroying a specific area of the brain by inserting tiny electrical wires in the area has shown promising results, but the search for better ways to treat Parkinson's disease is still ongoing.

Schizophrenia is a severe mental disorder in which patients' judgment is impaired and they lose contact with reality. Schizophrenics' symptoms include thought disorders, in which thoughts and speech are disconnected; delusions and hallucinations; mood disorders, including depression, anxiety, and euphoria; restlessness or inactivity; and withdrawal from social interactions. Two observations suggest that aspects of schizophrenia may be due to alterations in dopamine synaptic transmission.

First, the amphetamines that raise brain dopamine levels can induce in humans a state that closely resembles schizophrenia. Second, the standard treatments for schizophrenia are drugs that block dopamine receptors; these drugs prevent dopamine from activating dopamine receptors. The most commonly prescribed agents are haloperidol and clozapine, and they are quite effective in treating many symptoms, especially delusions and hallucinations. For many schizophrenics these antipsychotic drugs have been a godsend because they enable them to function reasonably well in society.

Although indirect evidence suggests that the underlying cause of schizophrenia involves dopamine synapses, we really have no direct evidence. As yet, no study has provided unequivocal evidence for altered dopamine neurotransmission in the brains of patients suffering from schizophrenia. No region of the brain has been found, for example, that possesses excess levels of dopamine.

A problem with treating any brain disease with drugs is side effects. Schizophrenics are given drugs that block dopamine action; thus, the presumption is that their brains contain excess dopamine. Parkinson's disease, in contrast, is due to a deficit of dopamine. Do schizophrenics treated with dopamine-blocking agents develop Parkinson-like symptoms? They often do, which limits the amounts and types of dopamine-blocking agents that can be given to schizophrenics. Conversely, some people with Parkinson's disease develop schizophrenic-like symptoms when given L-dopa, presumably because dopamine levels are increased in some region of the brain. For certain patients, the schizophrenic-like symptoms are so severe that L-dopa therapy must be discontinued. Withdrawing the L-dopa therapy reduces the schizophrenia-like symptoms, but then Parkinsonian symptoms reappear.

Serotonin and Depression

Most of us have experienced a bout with depression. For some people, depression can become prolonged and debilitating, as was the case for Tess and her mother described at the beginning of this chapter. The possibility that depression may be related to an upset in monoamine synaptic transmission came from observations in the 1950s when patients were treated with two kinds of drugs for reasons totally unrelated to brain function. Reserpine, which treats hypertension, caused depression in many individuals. Reserpine was subsequently found to deplete monoamines in nerve terminals, so it was supposed that depression may relate to depressed levels of monoamines. By contrast, drugs for treating tuberculosis (iproniazid and isoniazid) often alleviated depression. These anti-tubercular drugs are inhibitors of the enzyme monoamine oxidase that breaks down monoamines in the brain. In patients treated with iproniazid or isoniazid, therefore, monoamine levels are elevated.

These early observations led pharmacologists to develop and test several monoamine oxidase inhibitors as possible antidepressants, and many of them did show antidepressant activity. But side effects and low efficacy in patients with severe depression led pharmacologists to seek drugs that raise monoamine levels in a different way, by inhibiting the reuptake of monoamines into nerve terminals. The resulting compounds, called *tricyclics*, were more successful in treating depression than are monoamine inhibitors. Like the monoamine inhibitors, however, the early tricyclic drugs were not very specific with regard to monoamines; all the monoamines were affected, as were other neuroactive substances, and this often led to unpleasant side effects.

The more effective tricyclics inhibited the uptake of just two monoamines, serotonin and norepinephrine, and so a third generation of antidepressant drugs was developed to be more specific in the monoamine reuptake they inhibit. One of these, Prozac (fluoxetine), potently and selectively inhibits serotonin uptake by inhibiting the transporter present at synapses releasing serotonin. Thus, it specifically increases serotonin levels at these synapses. This drug and related ones have been quite successful; they are effective with all kinds of people and with a number of affective disorders aside from depression. These drugs have few side effects and are tolerated well by most patients. Prozac was the drug that caused the amazing transformation of Tess, described at the beginning of this chapter.

The lesson here is that the more specific a drug is in relation to the molecules it affects at a synapse, the more successful it may be in treating a disorder and the fewer side effects it may have. For example, the receptors that interact with a specific neuromodulator (or neurotransmitter) are not all identical. Typically there are several subtypes of receptors that differ slightly in their protein structure. At any one synapse, usually only one receptor subtype is present. Receptor subtypes can be distinguished pharmacologically, and one challenge for pharmacologists is to find compounds that interact with a specific subtype of receptor. The new drugs that combat schizophrenia primarily block the effect of dopamine on only one of the five known dopamine receptors. These newer drugs, such as clozapine, cause fewer Parkinson-like side effects than the less specific drugs, such as haloperidol.

Neuropeptides: The Enkephalins—Endogenous Opiates

By far the most diverse group of substances released at synapses are the neuropeptides. They are also the least understood in terms of their mode of action, but it is believed that neuropeptides generally act as neuromodulators. Alterations in brain levels of several neuropeptides have been linked to mood disorders. Some of these peptides interact with the dopaminergic and serotonergic systems in the brain, and the mood disorders may be accounted for by their effects on monoamine synaptic transmission. It may also be the case that certain peptides themselves play a central role in affective states of the brain.

As many as thirty different *peptides* are released at brain synapses. These peptides vary widely in size, containing between three and forty amino acids. They are commonly divided into four groups, depending on where they were first identified. For example, two groups, the hypothalamic peptides and pituitary peptides, participate in the release of hormones from the hypothalamus, pituitary, or other glands. A third group was identified first in the digestive system, where they are responsible for regulating aspects of digestion. Finally, some peptides, the *enkephalins*, were first identified in brain tissue. Figure 3.4 lists representative examples for each of these four groupings.

In any region of the brain the number of neurons containing and releasing a peptide from its terminals is small. The peptide-containing neurons, though, trigger wide-ranging processes, which means that their roles are more global. But as yet, specific roles for most peptides in brain function have not been established. A striking exception, though, are the enkephalins, which appear to act as endogenous opiates. Opiates such as morphine, the active ingredient of the opium poppy, have long been used to relieve pain and to treat medical problems ranging from coughing to diarrhea. The opiates also cause pleasant mental effects and are classic recreational drugs. It is not news that the opiates are highly addicting and have presented a serious problem around the world for centuries. After the Opium Wars of the early nineteenth century, when the British forced the Chinese to open their markets to opium and permitted the massive importation of opium into China, a fourth of the Chinese population became addicted to opium. Today, society has the same problem with heroin and other synthetic opioids, which have been chemically modified and are more potent than natural opioids.

The discovery of enkephalins was prompted by hints that receptors for opiates might reside within the brain. Why? For two reasons. The first was that many opiates are effective at tiny concentrations, which suggested the possibility of specific recognition sites for such molecules in the brain. The second was the discovery of substances that block the

Hypothalamic peptides

THYROTROPIN-RELEASING HORMONE (TRH)
Causes the pituitary gland to release thyrotropin

Glu - His - Pro

SOMATOSTATIN
Inhibits the release of thyrotropin and growth hormone from the pituitary gland

Ala - Gly - Cys - Lys - Asn - Phe - Phe - Trp
Cys - Ser - Thr - Phe - Thr - Lys

LEUTEINIZING HORMONE-RELEASING HORMONE (LHRH)
Promotes the release of leuteinizing hormone from the pituitary

Glu - His - Trp - Ser - Tyr - Gly - Leu - Arg - Pro - Gly

Pituitary Peptides

VASOPRESSIN
Causes reabsorption of water by the kidney and blood vessel constriction

Phe - Tyr - Cys
Gln - Asn - Cys - Pro - Arg - Gly

CORTICOTROPIN (ACTH)
Causes release of steroid hormones from the adrenal glands

Ser - Tyr - Ser - Met - Glu - His - Phe - Arg - Tyr - Gly - Lys - Pro - Val - Gly - Lys - Lys - Arg - Arg - Pro - Val - Lys - Val - Tyr - Pro - Asp - Gly - Ala - Glu - Asp - Glu - Leu - Ala - Glu - Ala - Phe - Pro - Leu - Glu - Phe

Digestive system peptides

CHOLECYSTOKININ (CCK)
Initiates the release of bile from the gall bladder

Asp - Tyr - Met - Gly - Trp - Met - Asp - Phe

VASOACTIVE INTESTINAL PEPTIDE (VIP)
Causes constiction of blood vessels in the intestine

His - Ser - Asp - Ala - Val - Phe - Thr - Asp - Asn - Tyr - Thr - Arg - Leu - Arg - Lys - Gln - Met - Ala - Val - Lys - Lys - Tyr - Leu - Asn - Ser - Ile - Leu - Asn

SUBSTANCE P
Causes the contraction of smooth muscle in the digestive tract

Arg - Pro - Lys - Pro - Gln - Gln - Phe - Phe - Gly - Leu - Met

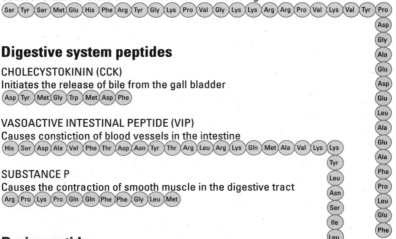

Brain peptide

MET-ENKEPHALIN

Tyr - Gly - Gly - Phe - Met

FIGURE 3.4 *Representative neuropeptides found in the brain. Many of these neuropeptides are involved in endocrine function and were first identified in the hypothalamus, pituitary gland, or digestive system. Their roles in endocrine function are noted here, along with their amino acid structures. Met-enkephalin is found only in the brain. Amino acid abbreviations: Ala, alanine; Arg, arginine; Asn, asparagine; Asp, aspartic acid; Cys, cysteine; Gln, glutamine; Glu, glutamic acid; Gly, glycine; His, histidine; Ile, isoleucine; Leu, leucine; Lys, lysine; Met, methionine; Phe, phenylalanine; Pro, proline; Ser, serine; Thr, threonine; Trp, tryptophan; Tyr, tyrosine; Val, valine.*

effects of opiates. One such drug, naloxone, seemed to act like other substances that prevent a neurotransmitter or neuromodulator from interacting with its channel or receptor.

In the 1970s, Candace Pert and Solomon Snyder at Johns Hopkins University proved that brain tissue has specific receptors for opiates; these findings suggested that natural opiate-like substances occur in the brain. Such substances were looked for in brain tissue, and shortly thereafter, two small peptides that bind to opiate receptors, the enkephalins, were isolated from brain tissue.

What role might the enkephalins play in brain function? The answer remains elusive, but their discovery explains some puzzling aspects of pain and how it can be relieved in unconventional ways. For example, the relief of pain by acupuncture treatment probably results from the release of enkephalins within the brain. Evidence that this is so comes from experiments on animals, which were shown to experience higher pain threshold levels than controls when given acupuncture treatment. However, if the animals were first given naloxone, which presumably blocked the opiate receptors in their brains, acupuncture treatment had no effect on pain thresholds.

The placebo effect that results in pain relief is another phenomenon that may be explained by the endogenous release of opiate-like substances. A convincing experiment was conducted at the University of California San Francisco Medical Center by Howard Fields and his colleagues. Medical students who had had a wisdom tooth extracted were given either morphine or a sugar pill *placebo*. They were told these therapies would decrease pain, and both groups experienced less pain than a control group given nothing. When a subset of the students given the placebo was also given naloxone, these students experienced pain levels comparable to the controls given nothing. The conclusion drawn was that a pill thought to relieve pain released an enkephalin or enkephalin-like substances in the brain.

Of what value would opiates released within the brain be? Soldiers severely wounded in battle and athletes hurt in competition often feel little pain until the battle or contest is over. Why? The release of endogenous opiates could explain the lack of pain. Even strenuous activity appears to release endogenous opiate-like substances, as shown by the block by naloxone of the "high" often experienced by long-distance runners. One can envision the survival or merciful value of such substances released in the brains of wounded humans or animals that are being attacked or chased. A vivid description of how powerful this system may be was provided by David Livingstone, the Scottish missionary and explorer who was attacked by a wounded lion in Africa. Fortunately for Livingstone, the lion was quickly distracted by another member of his party who was attempting to shoot the beast. The lion left Livingstone to attack the gunman. A third man was also bitten before the lion fell dead from his wounds. All three survived the lion's attacks. Livingstone wrote of his reactions to the attack in his *Missionary Travels* (London: John Murray, 1875):

> I heard a shout. Starting, and looking half around, I saw the lion just in the act of springing upon me. I was upon a little height, he caught my shoulder as he sprung, and we both came to the ground below together. Growling horribly close to my ear, he shook me as a terrier does a rat. The shock produced a stupor similar to that which seems to be felt by a mouse after the first shake of the cat. It caused a sort of dreaminess in which there was no sense of pain nor feeling of terror, though quite conscious of all that was happening. It was like what patients partially under the influence of chloroform describe, who see all the operation, but feel not the knife. . . . The shake annihilated fear, and allowed no sense of horror in looking round at the beast. This peculiar state is probably produced in all animals killed by the carnivora; and if so, is a merciful provision by our benevolent creator for lessening the pain of death.

Other receptors and endogenous substances may mediate similar types of phenomena. For example, a receptor for marijuana and marijuana-like substances has been identified in brain tissue, and endogenous substances that bind to these receptors have been found. What role such receptors or substances may play is not known.

To conclude, despite all that is still mysterious about brain chemistry, drug therapy for mental disease has been a tremendous success story.

Many severely disturbed patients previously doomed to a life in a mental institution can now function in society as a result of drug therapy, and less debilitating mental disturbances such as anxiety and depression can often be relieved quite effectively with a variety of substances. Drug therapy is by no means a panacea; drugs treat symptoms and don't cure the disease. For many, side effects are a problem, and in some cases, drug therapy fails. Nevertheless, drug therapy has been an enormous advance for the treatment of mental disease, and as we learn more of the intricacies of brain function, especially synaptic mechanisms, new, more effective, and more specific agents will undoubtedly be found.

Prior to drug therapy, the most effective treatment for mental illness was, of course, psychotherapy. But psychotherapy was not particularly effective with severely ill patients, and it is slow and expensive. The great success with drug therapy naturally raises the question of how psychotherapy affects the brain. There is evidence that psychotherapy can alter brain chemistry, so the two treatments may be more complementary than is intuitively obvious. How might psychotherapy alter brain chemistry? The placebo phenomenon described above may provide a model. That is, if a subject believes a pill will relieve pain, it does so even if the pill is just sugar. Ingestion of the placebo causes the release of endogenous opiate-like substances in the brain—brain chemistry is altered and profound effects can result. It is quite conceivable that a similar type of alteration in brain chemistry—an increase or decrease in the release of a neurotransmitter or neuromodulator—occurs as a result of psychotherapy. Indeed, most psychiatrists today view psychotherapy as an essential adjunct of drug therapy. They believe drug therapy is more effective when combined with psychotherapy. In Chapter 9 I describe brain plasticity and the idea that everything we do or experience can change the brain to some extent.

4

Sensing the World

Without our sensory receptors we would be isolated from the world and each other. Our five sensory systems respond to specific stimuli—light, sound, touch, airborne chemicals (smell), and solubilized molecules (taste). The first three are the most important for us to interact with the world and with each other, as the story of Helen Keller illustrates.

At 19 months of age, Helen lost both vision and hearing because of a devastating illness. As Helen grew, she was often depressed, and even angry, having frequent temper tantrums. This changed when her teacher, Annie Sullivan, hired to work with Helen when she was seven, managed to communicate with her through the sense of touch. Annie first taught Helen the manual finger alphabet by tapping on the palm of her hand. Initially, Helen had little understanding of what the various letter taps meant, but the breakthrough came one day when Annie and Helen were in the family pump house. As water from the spout rushed over one of Helen's hands, Annie spelled w-a-t-e-r in the palm of the other hand. The sensation of water rushing over one hand, feeling the letters spelled out in the other, was transforming for Helen; she instantly understood and wanted to know the names of everything she touched. In a few hours, she learned the names of thirty objects. Within 4 months, Helen had a vocabulary of 400 words. She learned language much as a child learns spoken language, but she did so amazingly fast. By the end of their first year together, Annie was spelling into Annie's hand stories from the *Iliad* and *Odyssey*.

Together, Helen and Annie achieved remarkable things, including Helen earning a college degree from Harvard's sister school Radcliffe in 1904, the first blind and deaf person ever to earn a college degree.

—Adapted from Dorothy Herman, *Helen Keller* (Chicago, IL: University of Chicago. Press, 1999)

All neurons have characteristics of sensory receptors in that they possess specialized membrane proteins that respond selectively to specific chemicals—the neurotransmitters or neuromodulators. The effects of neurotransmitters on neurons, directly producing potential changes across the cells' membrane, are similar to the effects of stimuli on touch, auditory, and some taste receptors. Olfactory receptors, photoreceptors, and certain taste receptors respond to stimuli as neurons respond to neuromodulators: enzyme cascades are activated, resulting in alterations of second-messenger levels and excitation of the cell.

This chapter describes examples of these two basic types of receptors: directly gated receptor cells and receptor cells gated by second messengers. In directly gated receptors, such as touch receptors, deformation of membrane channels themselves or of surrounding membrane results directly in the opening of the channels and in the generation of receptor potentials. Activation of olfactory or photoreceptors, in contrast, leads to activation of enzyme cascades by membrane receptors and to changes in second-messenger levels. Alterations in second-messenger levels result in the generation of receptor potentials.

Mechanoreceptors: Touch and Hearing

A great variety of receptors found in the skin and other tissues respond to mechanical stimuli, including touch, pressure, and vibration. Some *mech-*

anoreceptors are even found in muscles, and they signal muscle stretching and tension. Our model is the *Pacinian corpuscle*, a large pressure receptor found in the skin, muscle, joints, and tendons. The Pacinian corpuscle has been studied extensively because of its large size, so we know more about it than any other touch or pressure receptor, all of which are thought to work much like the Pacinian corpuscle. Under the microscope the corpuscle looks like a slice of onion, with many concentric layers of flattened nonneural (epithelial) cells surrounding a bare nerve ending. The nerve ending becomes myelinated upon emerging from the corpuscle. This corpuscle and a length of nerve fiber can easily be excised from tissue and kept functioning for some time; hence, they are convenient for study.

Figure 4.1 also shows what a Pacinian corpuscle receptor cell looks

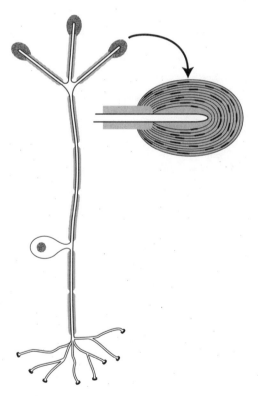

FIGURE 4.1 *A typical pressure receptor. The neuron is monopolar; a short process from the cell body divides into two, with one myelinated process extending to the sensory endings (Pacinian corpuscles) and the other to the cell terminals. A Pacinian corpuscle (enlargement on right) consists of a bare nerve ending surrounded by a capsule of flattened nonneural cells.*

like. It is monopolar in nature—that is, only one process comes from the cell body. This process divides into two myelinated processes: one goes to the sensory endings, and the other to the spinal cord. Action potentials generated in the sensory endings are propagated directly to terminals in the spinal cord. The cell bodies of sensory cells innervating the limbs and trunk lie just adjacent to the spinal cord in structures called dorsal root ganglia. With a preparation consisting of the corpuscle and a short length of nerve hooked with a wire, it is easy to stimulate neural activity with a fine needle or stylus (see Figure 4.2). Compression of the corpuscle by a tiny amount, just a fraction of a micrometer (0.2–0.5 µm), induces a small receptor potential. With more compression, provided by a stronger stimulus, a

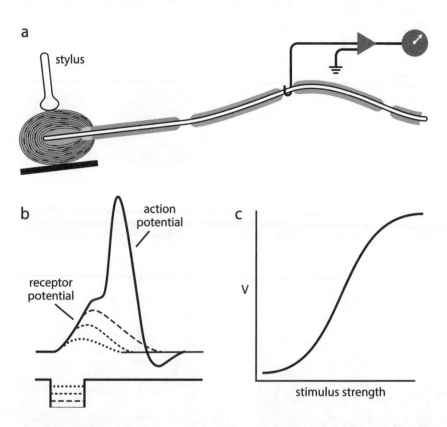

FIGURE 4.2 When a Pacinian capsule is depressed with a stylus (a), depolarizing receptor potentials are evoked in the nerve that, if large enough, generate an action potential (b). (c) An S-shaped relation typically exists between receptor voltage (V) and stimulus strength.

larger receptor potential is generated. A depolarization of 10–15 millivolts or so generates an action potential superimposed on the receptor potential.

Action potentials can be blocked by the drug *tetrodotoxin*, but the receptor potential is not. This is because tetrodotoxin blocks voltage-gated Na⁺ channels but not channels underlying receptor or synaptic potentials. By bathing the preparation in a solution containing tetrodotoxin, one can study the receptor potential in isolation, without the intrusion of action potentials. Intensity-response relations for the receptor potential can then be worked out (e.g., Figure 4.2c). Many types of receptors have similar S-shaped intensity-response relations. How is this receptor potential generated? Compression of the myelinated fiber coming from the capsule yields no response unless a large distortion of the membrane (10–15 micrometers) is induced. Then a change in membrane potential is recorded that probably represents damage to the nerve. (This is analogous, probably, to the excitation of the ulnar nerve when you hit your elbow on a hard object, that is, when you hit your crazy or funny bone.) Removing the onion-like capsule does not affect the response either. Indeed, when a preparation is stripped of the capsule, small deformations all over the bare nerve ending generate small receptor potentials. These small receptor potentials summate and produce a larger receptor potential. If the first node of the myelinated part of the fiber is mechanically blocked (by applying pressure to the node), no action potentials are generated but receptor potentials are still recorded.

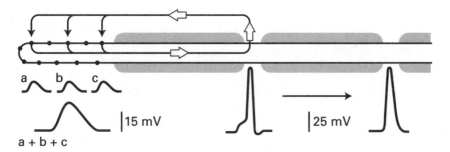

FIGURE 4.3 *When the Pacinian capsule is removed, leaving a bare nerve ending, deformations of the membrane (arrows) result in small depolarizing potentials (a, b, and c) that can summate. Current (open arrows) entering the fiber through the mechanosensitive channels (filled circles) results in depolarization of the membrane at the first node, where voltage-gated channels are present, resulting in the generation of action potentials, which then travel down the fiber.*

The experimental results suggest the following model. Contained within the membrane of the bare nerve ending are specialized channels whose conductance is altered when they or the surrounding membrane is deformed. At rest, few if any of the channels are open, but when the membrane is stretched, the channels change configuration and allow ions to cross the membrane. Because of the conductance change, net positive charge accumulates inside the cell, depolarizing the membrane at the first node in the fiber's myelinated part. Here voltage-sensitive channels are located, and with sufficient depolarization, action potentials are generated that propagate down the fiber. Figure 4.3 is a diagram of the Pacinian corpuscle based on this model. Positive ions enter the bare nerve ending where the membrane is deformed and flow internally down the fiber. This depolarizes the first node in the fiber's myelinated part and generates action potentials that travel down the cell.

What ions generate the receptor potential? This question can be answered by altering ion concentrations in the extracellular solution. Changes of Na^+ most affect the receptor response; hence, the channels are mainly permeable to Na^+, as is the case for channels that give rise to excitatory synaptic potentials (see Figure 2.9).

Adaptation

The Pacinian corpuscle also dramatically illustrates an important property of all receptors: *adaptation*. With a sustained stimulus, all receptor potentials decline. They may do so rapidly or slowly, completely or partially. Sensory adaptation is a distortion of the real world in the sense that receptors do not provide a faithful representation of the stimuli impinging on the organism. Even so, sensory receptor adaptation can have significant advantages; it is unlikely, for example, that we could wear clothes without being severely distracted by the continuous response of our touch receptors if they did not rapidly adapt.

Three types of adapting receptors are distinguished: fast, medium, and slow. Fast-adapting or phasic receptors respond only to a change in stimulus level. Pacinian corpuscles are fast-adapting receptors, as are other touch receptors and olfactory receptors. Figure 4.4a shows the response

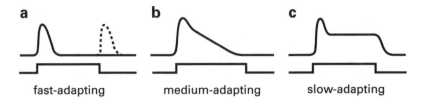

FIGURE 4.4 *Sensory receptor adaptation. (a) Fast-adapting receptors respond only for a short time after a stimulus is applied. With fast-adapting receptors, a response at stimulus offset following a prolonged stimulus is often observed. (b) Medium-adapting receptor also decline in response amplitude over time, but more slowly. (c) Slow-adapting receptors show an initial decline in response amplitude, but then plateau and the response lasts as long as the stimulus is maintained.*

of a Pacinian corpuscle to a prolonged stimulus. Upon application of pressure to the receptor it rapidly depolarizes, but within 10 milliseconds or so the voltage decays to zero even though the pressure is maintained. The Pacinian corpuscle displays another feature of many phasic receptors in that it responds at the offset of the stimulus as well as at its onset.

Medium-adapting receptors decline much more gradually in potential during the presentation of a long stimulus. If the stimulus is sufficiently prolonged, the response can decay almost completely to baseline, as illustrated in Figure 4.4b. Taste and hearing receptors are of this type. Photoreceptors and deep pressure receptors are slow-adapting or tonic receptors (Figure 4.4c). Photoreceptors respond to a prolonged stimulus in two stages, an initial transient potential that decays to a smaller plateau potential, which is maintained for as long as the stimulus continues. Although these receptors respond for as long as the stimulus is applied, as with other receptors they give a maximal response at stimulus onset and a smaller response thereafter. All receptors, even tonic ones, respond best to changes of stimulus level.

Hair Cells and Audition

The sensory receptors for hearing are called *hair cells*, and they, too, are mechanoreceptors. Hair cells, like the Pacinian corpuscle, have channels that open when the channels themselves or the surrounding membrane are stretched or deformed. These channels are found at the tips of hairlike projections that extend from the top surface of the cell, and the hairs are connected directly by fine filaments that run between adjacent "hairs."

When sound impinges on the fluid-filled inner ear, as shown schematically in Figure 4.5, the hair cells move relative to the *tectorial membrane*

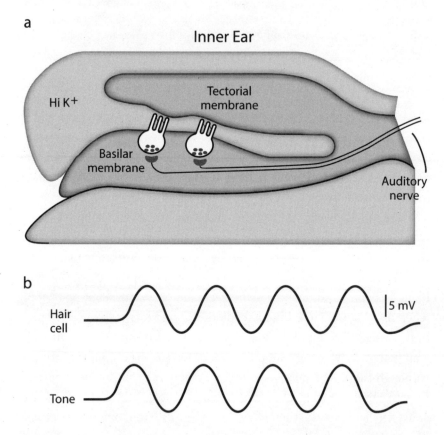

FIGURE 4.5 *(a) Representation of hair cells in the inner ear. When sound impinges on the ear, the tectoral membrane moves relative to the basilar membrane (see Figure 4.6b), causing the hairs on the cell to bend. This causes the fine filaments connecting the hairs to stretch, thus opening channels on the hairs, allowing K+ to flow into the cell and causing the cell to depolarize. (b) An intracellular recording from a hair cell shows a change in membrane potential that closely matches the tone stimulus. When the cell depolarizes, transmitter is released from the hair cell, activating the auditory nerves that contact the hair cells.*

to which the hairs are attached. The hairs bend, increasing tension on the filaments and causing the channels to open, which allows positive ions to cross the membrane. The fluid surrounding the hairs in the inner ear is high in K^+; indeed, K^+ is higher outside the hairs than inside. Thus, K^+ flows into the hair cell and depolarizes it when the channels are opened.

Action potentials are not generated by hair cells. Rather, the graded receptor potential leads to the release of transmitter from the cell's synapses; it is the second-order cells that first generate action potentials in the auditory system. A similar situation is found in the visual system. Indeed, in the vertebrate retina, action potentials are first generated mainly by third-order cells. That is, in the vertebrate retina excitation and synapse activation is via graded potentials in both the receptor and the second-order cells.

Hair cells serving hearing are found in a complex structure, the cochlea, situated in the inner ear (Figure 4.6). The cochlea is a rigid, coiled structure, encased in bone, that gets narrower as it goes from base to apex—hence, it resembles a snail. Indeed, the word *cochlea* derives from the Greek word for snail. A cross section through the cochlea shows that it consists of three fluid-filled compartments (Figure 4.6b). The hair cells are in the central compartment, termed the scalia media, in a structure called the *organ of Corti* that sits in the *basilar membrane*. The other two compartments are in continuity with each other at the apex of the cochlea, whereas at the base of the cochlea they are in contact with two membrane-covered holes called the oval and round windows.

When sound impinges on the ear, the eardrum vibrates. Three small bones transmit the vibration to the oval membrane, which induces vibration of the fluid in the upper and lower chambers and of the round window. Both the tectorial and basilar membranes flex in response to the fluid movements but differentially, resulting in bending of the hairs on the hair cells; this is key to excitation of the hair cells.

If the cochlea were unrolled, it would be about 32 millimeters long. Although the cochlea narrows as it goes from base to apex, the opposite is the case for the basilar membrane: it is five times wider at the apex than at the base. Furthermore, its rigidity decreases from base to apex about a hundredfold. At the apex, the basilar membrane is much more flexible than at the base. The result is that the region of maximal vibration along

a

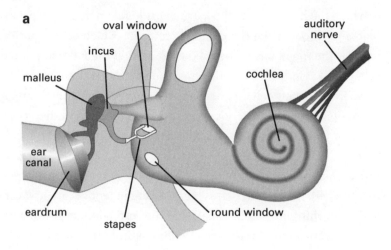

b

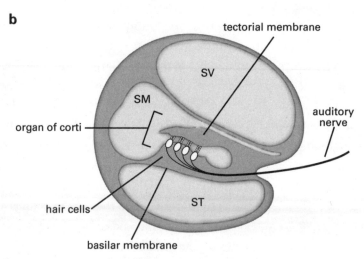

FIGURE 4.6 (a) The cochlea in the inner ear. Sound impinging on the ear drum causes vibrations that are transmitted to the oval window of the cochlea via three small bones: the malleus, incus, and stapes. (b) A cross section of the cochlea illustrating its three chambers: the scala vestibuli (SV), scala media (SM), and scala tympani (ST). The organ of Corti sits in the scala media and consists of hair cells, which are embedded in the basilar membrane and the overlying tectoral membrane. (See Figure 4.5 for a closer view.) The auditory nerve innervates the hair cells.

the basilar membrane depends on sound frequency. In this way place coding is established along the basilar membrane that determines pitch. In other words, we can detect pitch—different tones—by which hair cells along the basilar membrane are maximally activated.

Second-Messenger Receptors: Olfactory and Visual

The olfactory and taste systems have many features in common. Olfaction is specialized to respond to airborne molecules whereas taste detects solubilized molecules. A variety of taste receptors are known that respond to different-tasting substances. Salts and some amino acids interact directly with channels on the receptor cells, resulting in the depolarization and activation of the taste cells, much as excitatory neurotransmitters activate neurons. Other substances, including sugars and bitter compounds, activate G-protein-coupled receptor molecules on the receptor cells that link to either adenylate cyclase (sugars) or other synthetic enzymes (bitter substances). Depolarization of these taste receptor cells results ultimately from activation of G-protein-related molecules. All olfactory receptors appear to be activated via second-messenger pathways, and they will be our model. In some olfactory receptor cells a cAMP pathway is activated, whereas in others a different second messenger system is activated. Different types of odors preferentially activate one or the other pathway. Fruity odors, for example, such as essence of lemon, activate primarily the cAMP pathway, whereas offensive odors, such as those found in sweat, activate another pathway.

The olfactory receptors are found in the olfactory epithelium that lines the interior of the nose (see Figure 4.7a). The receptor cells are characterized by cilia that extend from a prominent apical dendrite into a thick layer of mucus that lines the nasal cavity. The mucus is produced by cells and glands that sit under the olfactory epithelium. Supporting cells are also present in the epithelium. Odorants entering the nose become dissolved in the mucus and are detected by their presence in the mucosa.

The odorant receptor molecules are found on the cilia. Application of an odorant to the cilia will cause a depolarization of the receptor cell, whereas odorant applied to the cell body causes little or no response. Intracellular recordings from olfactory receptors show that in response

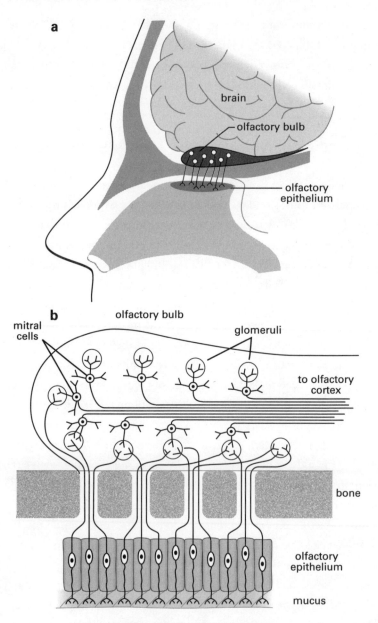

FIGURE 4.7 (a) *The olfactory epithelium is located in the nasal cavity just below the olfactory bulb. (b) The olfactory receptors, which lie in the olfactory epithelium, extend their axons into glomeruli in the olfactory bulb, where they innervate the dendrites of mitral cells. The mitral cells carry the olfactory signal to the olfactory cortex. Each glomerulus is odor specific; that is, receptors that express the same odorant receptor molecules project to the same glomerulus.*

to odorants applied to the cilia, a very large depolarizing receptor potential, up to 50 millivolts in amplitude, is generated that passively spreads down the dendrite and into the cell body. Action potentials are generated where the axon comes off the cell and propagate along the olfactory nerve axons into the olfactory bulbs, which sit above the nasal cavity below the brain (Figure 4.7b). The olfactory nerve axons end in specialized spherical structures called *glomeruli*, which I discuss further below.

The way olfactory receptor cells are activated is as follows. An odorant molecule activates a G-protein-related receptor molecule that activates either adenylate cyclase or another synthetic enzyme, resulting in increased levels of cAMP (Figure 4.8) or another second messenger within the cell. cAMP interacts directly with a channel in the membrane—no phosphorylation

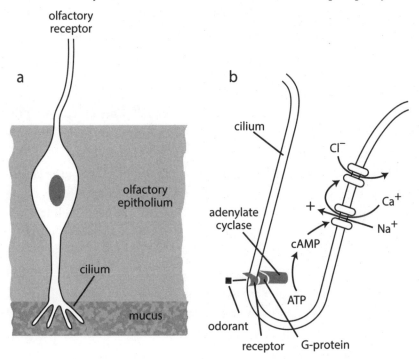

FIGURE 4.8 *(a) An olfactory receptor cell. Cilia extend from an apical dendrite on the receptor cell into a mucus layer that lines the nasal cavity. (b) Olfactory transduction occurs in the cilia. An odorant interacts with a receptor linked to a G-protein, which in turn is linked to an enzyme (in this case, adenylate cyclase). The cAMP generated by adenylate cyclase opens channels in the membrane that allow Na^+ and Ca^{2+} to enter the cell. The Na^+ depolarizes the cell, whereas the Ca^{2+} activates a Cl^- channel, which promotes Cl^- efflux from the cell, leading to further depolarization.*

is involved—that allows both Na⁺ and Ca²⁺ ions to cross the membrane. The influx of Na^+ depolarizes the cell, whereas the Ca^{2+} ions (which by themselves will also cause some depolarization) interact with a membrane channel that permits Cl^- to leave the cell. Cl^- leaving the cell adds to the depolarization by reducing the negative charge within the cell, and thus the Ca^{2+}-activated Cl^- current acts as an amplifying mechanism, explaining the large receptor potentials that can be generated in olfactory receptor cells.

Discrimination of Odors

Humans can discriminate as many as 10,000 odors. How does this happen? It turns out that many different odorant receptor molecules are found in the olfactory epithelium, almost four hundred in humans and two to three times more in other animals. Each of these receptor molecules is coded by a separate but related gene; they make up the largest family of related genes known. Furthermore, it appears that only one or a few of these genes are expressed in a particular olfactory receptor cell, which means that the olfactory epithelium has a large number of distinct receptor cells. How are the coding and discrimination of different odors managed?

One receptor cell will respond to more than one odor, but each cell responds best to a specific odor. In other words, olfactory receptors have "tuning curves"—they respond best to a specific odor, less well to other odors. In this way they are similar to hair and photoreceptor cells, which respond maximally to one frequency of sound or wavelength of light and less well to others. The surprise in the olfactory system is how many different receptors there are; in contrast, in the visual system only three or four different receptors are required to discriminate all the colors in the spectrum.

How are the responses of the various olfactory neurons sorted out? Receptor cells of the same type—that express the same odorant receptor molecules—innervate separate structures (termed glomeruli) in the olfactory bulb (Figure 4.7b). In mice, each glomerulus contains the dendrites of up to one hundred glomerular (mitral) cells. Impinging on these dendrites are as many as 25,000 olfactory receptor axons, all of the same type. The *mitral cell* axons from a single glomerulus thus send to the rest of the brain quite specific odorant information.

Visual Receptors

Most vertebrate eyes have two types of light-sensitive photoreceptor cells, called *rod photoreceptors* and *cone photoreceptors*. Rods mediate dim-light vision, whereas cones function in brighter light and mediate color vision. Usually there is just one type of rod but several types of cones in vertebrate eyes. Humans, for example, have three types of cones: one that responds best to red-yellow light, another to green light, and a third to blue light. Rods and cones are elongated cells with a specialized outer segment region and an inner segment and synaptic terminal (Figure 4.9a). The outer segments consist of numerous membrane infoldings, usually pinched off from the outer membrane in rods, but still connected to the outer membrane in cones.

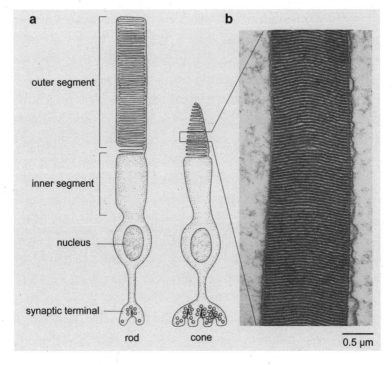

FIGURE 4.9 *(a) In both rod and cone photoreceptors, light-sensitive visual pigments are found in membranous disks in the outer segments of the cells. (b) A portion of cone outer segment of a lizard. The lizard cone outer segments are long, and over a short portion of their length the cone shape is not obvious.*

Photoreceptors "see" because they contain abundant light-sensitive molecules, the *visual pigments*, in the membranes of the cell's outer segment (Figure 4.9b). Light is captured (absorbed) by these molecules, and this leads to excitation of the photoreceptor cell. The light-sensitive molecules are called pigments because they absorb certain wavelengths of visible light and hence have color. The visual pigment in rods, called *rhodopsin*, absorbs blue-green light best and captures red and blue light less well; because it lets red and blue light escape, it appears purple to us. If one removes the retina from an eye of a dark-adapted animal that contains abundant rods (and most animals, including ourselves, have many more rods than cones), the entire retina has a reddish purple hue. Indeed, the original name for the rod pigment was visual purple.

Light does two things when captured by the visual pigment molecule. First, it activates the molecule and excites the cell; second, it breaks down the molecule into component parts. All visual pigments consist of a large protein to which is bound a slightly modified form of vitamin A: vitamin A aldehyde, or *retinal*. When bound together, the molecule is sensitive to light in the visible range of the spectrum—between deep blue and far red light or between 400 and 700 nanometers. When split, the two components of the molecule absorb mainly ultraviolet light, which is invisible to us. The breakdown or bleaching of the visual pigment molecules in the light inactivates or desensitizes the photoreceptors. So when you go from bright light into a dark theater, it takes many minutes for your eyes to adjust. What you are waiting for is the resynthesis of the visual pigments in the photoreceptor cells—a process known as *dark adaptation*—which restores the light sensitivity of the photoreceptors. After bright-light adaptation, cones dark-adapt in 5–6 minutes, but rods require up to 30 minutes to complete dark adaptation.

Individuals deficient in vitamin A are less sensitive to light than are well-nourished individuals. This condition is known as *night blindness* because it is most obvious at night. Yet both rods and cones are affected in vitamin A deficiency—both have visual pigments that require vitamin A, and thus with vitamin A deficiency both are less sensitive to light. What differs between the rod visual pigment and the three cone pigments are the proteins. Differences in these proteins give the visual pigment molecules somewhat different properties, including their color sensitivity, that is, the wavelength of light they best absorb.

The genes coding for the rod and cone visual pigment proteins thus are different. Defects or alterations in these genes lead to significant visual abnormalities. Color-blind individuals either have lost a gene or have a defective gene for one or another of the cone visual pigment proteins. Red-blind individuals are missing or have an altered red-sensitive visual pigment gene, green-blind individuals are missing or have an altered green-sensitive pigment gene, and blue-blind individuals are missing or have an altered blue pigment gene. The genes for the red- and green-sensitive pigments are on the *X-chromosome*. Since males have just one X-chromosome and females have two X-chromosomes, red and green color blindness is much more common in males than in females. This is because even with only one good chromosome, color vision will be normal; the photoreceptor cell can still make a normal pigment. Since females have two X-chromosomes, they can have one defective gene and one good gene and have normal color vision. Because males have only one X-chromosome, if that chromosome has a defective color pigment gene, the individual will be color-blind. Red-green color blindness is thus described as sex-linked.

It is important to note that most individuals we call color-blind still see colors. That is, if a person is unable to make one of the three cone visual pigments because of a defective gene (which is by far the most common situation), he or she still has two other cone types—cones sensitive to green and blue, red and blue, or red and green. With two cone types, color discriminations can be made, although such individuals cannot distinguish colors as well as a normal person with all three cone types. The color blindness exhibited by Jonathan I., described at the beginning of Chapter 7, was caused by a deficit in color vision processing in the brain, not by an alteration in his cones. He was totally color-blind; he could make no color discriminations at all.

Alterations in the gene for the rod pigment, rhodopsin, can lead to a disease called *retinitis pigmentosa*. People with this disease start off life with normal vision, but then the rods degenerate, first in the periphery of the retina and eventually throughout the retina. As the rods die, patients first lose the ability to see well in dim light, but eventually the cones die too, for reasons unknown, and all vision is lost. Why the rods die gradually, over the course of many years, is also not known. People with this genetic

disorder usually begin to notice diminished visual sensitivity in their twen-
ties or thirties. Complete blindness may come in the fifties or sixties.

Phototransduction

When a photon of light is absorbed by one of these light-sensitive molecules in a photoreceptor, it initiates a series of chemical reactions termed
phototransduction, (Figure 4.10). A light-activated visual pigment molecule
first interacts with and activates a G-protein (called *transducin*) that then
activates an enzyme called *phosphodiesterase (PDE)*. The second messenger in vertebrate photoreceptors is, however, cyclic guanosine monophosphate (cGMP), a molecule similar to cAMP but with guanosine rather
adenosine in its structure. Interestingly, cGMP levels are maintained at
a high level in vertebrate photoreceptors when in the dark. The cGMP
interacts with channels in the membrane that allow Na^+ and Ca^{2+} to cross
the membrane. The PDE activated by light breaks down the cGMP, thus
closing the membrane channels, resulting in the cell's membrane potential
becoming more negative—the cell hyperpolarizes, the opposite of what
happens in an olfactory receptor when it captures an odorant molecule. In
a sense, then, darkness is the stimulus for vertebrate photoreceptors, and
light turns them off. But it is the change of membrane potential in both
cases that is important and leads to excitation of the system.

Why do vertebrate photoreceptors behave this way? We do not know.
Another question is why do photoreceptors use a second-messenger system, and here we do have an answer: for amplification. That is, one activated visual pigment molecule can activate many transducin molecules,
and one transducin molecule can activate several PDE molecules. And an
activated PDE can break down many cGMP molecules. Thus, the amplification is considerable when a single visual pigment molecule absorbs a
photon of light, explaining the extraordinary sensitivity of a photoreceptor
responding to a single photon of light.

A final question is the role of Ca^{2+} in the process. As noted above,
the channels in the photoreceptors activated by cGMP allow both Na^+
and Ca^{2+} to enter the cell. Na^+ is much more abundant outside cells than
is Ca^{2+} and is mainly responsible for the change in membrane potential

when the cell absorbs a photon of light. Ca^{2+}, on the other hand, although present in lower amounts than Na^+, does inhibit the enzyme that makes cGMP. Thus, when the cGMP levels fall in the light, resulting in Na^+ and Ca^{2+} levels falling in the cell, the inhibition of cGMP synthesis also decreases, allowing an increase in cGMP levels, partially countering the effect of light. This results in a partial reopening of the Na^+/Ca^{2+} channels, explaining adaptation of the photoreceptor response This is why the photoreceptor response to light is initially larger than the subsequent sustained plateau responses even though the light impinging on the cell remains the same for the duration of the light stimulus. (see Figure 4.4c).

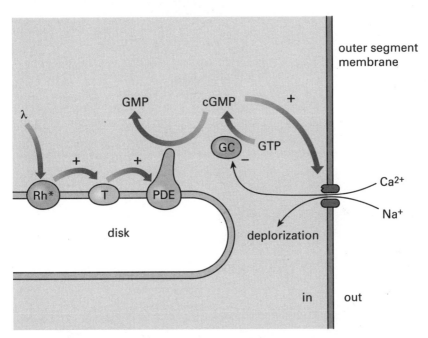

FIGURE 4.10 *A simplified scheme of the phototransduction process in the photoreceptor outer segment. Light-activated rhodopsin (Rh*) activates transducin (T) a G-protein, which in turn activates the enzyme phosphodiesterase (PDE). PDE breaks down cGMP to an inactive product (GMP); in the absence of cGMP, which opens channels in the outer-segment membrane, the channels close, and the cell hyperpolarizes (becomes more negative inside) because the positively charged Na^+ ions no longer can enter the cell. Also, Ca^{2+} levels in the cell decrease, allowing guanylyl cyclase (GC), normally inhibited by Ca^{2+}, to increase the synthesis of cGMP from guanosine triphosphate, an energy-rich molecule similar to ATP but with adenosine replaced by guanosine. With more cGMP available, more channels open, countering the effects of light. Ca^{2+} thus plays a role in photoreceptor adaptation.*

PART TWO

Systems Neuroscience

GETTING AT BEHAVIORS

Although a great deal has been learned about the processing of information by individual nerve cells, we are only beginning to understand the neurobiological basis of behavior. This is the realm of systems neuroscience—the study of aggregates, or networks, of neurons, and how they produce the range of behaviors we associate with brain function, from movement initiation and sensory processing to memory and learning. Both invertebrate and vertebrate nervous systems are being analyzed in complementary studies.

5

Simpler Nervous Systems:
THE INVERTEBRATES

The Marine Biological Laboratory (MBL) in Woods Hole is a paradigm, a human institution possessed of a life of its own, self-regenerating, touched all around by human meddle but constantly improved, embellished by it. . . .

Successive generations of people in bunches, never seeming very well organized, have been building the MBL since it was chartered in 1888. It actually started earlier, in 1871, when Woods Hole, Massachusetts, was selected for a Bureau of Fisheries Station and the news got round that all sorts of marine and estuarine life could be found here in the collisions between the Gulf Stream and northern currents offshore, plus birds to watch. Academic types drifted down from Boston, looked around, began explaining things to each other, and the place was off and running. . . . The giant axon of the Woods Hole squid became the apparatus for the creation of today's astonishing neurobiology. *Aplysia*, a sea slug that looks as though it couldn't be good for anything, has been found by neurophysiologists to be filled with truth. The invertebrate eye was invented into an optical instrument at the MBL, opening the way to modern visual physiology. Developmental and reproductive biology were recognized and defined as sciences here, beginning with sea-urchin eggs and working up. Marine models were essential in the early days of research on muscle structure and function, and research on muscle has become a major preoccupation at the MBL. Ecology was a sober, industrious science here long ago, decades before the rest of us discovered the term. In recent years there have been expansion and strengthening in new fields; biologic membranes, immunology, genetics, and cell regulatory mechanisms are currently booming.

—Excerpted from Lewis Thomas, *The Lives of a Cell: Notes of a Biology Watcher* (New York, NY: Penguin, 1995)

The nervous systems of animals without backbones—invertebrates—are simpler, with fewer neurons than are found in cats or catfish, frogs and finches, the vertebrates. Because of the relative simplicity of invertebrate nervous systems, neuroscientists have long studied them, and they have yielded some truly wonderful insights about nervous systems and brain function. The underlying assumption is that the same biological mechanisms operate in the nervous systems of all organisms—a not incorrect view. Invertebrates have provided invaluable knowledge about how single nerve cells function and how neuronal groups give rise to simple behaviors. Some of this research has revealed clues to higher brain phenomena like perception, learning, and memory.

The sea is a treasure-house of especially useful and accessible invertebrate organisms. The *squid*, with its giant nerve cell axon, has provided much of what we know about how nerve cells generate electrical signals. The *horseshoe crab* has an optic nerve that easily frays, and so it was first possible to record single optic nerve axons in that animal. Subsequent studies of the horseshoe crab visual system provided insights into how we perceive edges and borders. The sea slug *Aplysia* has a robust gill withdrawal *reflex*, one that is highly modifiable. This animal has elucidated how nerve cell function can be altered by experience—how a nervous system can learn and remember things. And much of this work was initiated and carried out at the MBL in Woods Hole, Massachusetts. More recently, behaviors have been analyzed by inducing gene mutations in animals and observing the behavioral effects of the altered genes. Fruit flies have been especially useful and yielding in this quest, and I end this chapter by describing how our understanding of the mechanisms underlying circadian rhythms derived from such research.

Electrical Signaling and the Squid Giant Axon

In vertebrates, as described in Chapter 2, most axons are covered by an insulting layer of myelin that accelerates action potential generation down the axon and streamlines its propagation. The glial cells of invertebrates, except in a few crabs, do not form myelin around nerve cell axons— invertebrate axons are without the benefit of myelin. An obvious consequence of the absence of myelin around invertebrate axons is that they conduct action potentials much more slowly than do vertebrate axons. While action potentials typically travel down vertebrate axons at rates of 100–200 miles per hour, they move along invertebrate axons at no more than 30–40 miles per hour. But even myelinated axons conduct action potentials incredibly slowly compared with how fast electrons flow down a conducting wire—at the speed of light!

Invertebrates can maximize how fast the action potential travels down axons by making their axons larger. When they do so, it lowers the axon's internal electrical resistance, which determines how easily ions can flow down the axon. Myelin, on the other hand, increases the cell membrane's electrical resistance, which governs how easily ions can flow across the membrane. What is critical for a high rate of action potential conduction down an axon is the ratio of membrane resistance to internal resistance. The higher the ratio, the faster is action potential conductance. So, decreasing an axon's internal resistance or increasing an axon's membrane resistance does very much the same thing biophysically: it speeds conduction. Why is this so? By decreasing internal resistance or increasing membrane resistance, the Na^+ ions admitted during an action potential can flow farther down the axon to depolarize the membrane and initiate the generation of new action potentials (see Chapter 2). But enlarging axon size to raise its conduction speed has limitations. Axons can be made only so big, and even the largest invertebrate axons conduct action potentials more slowly than the great majority of myelinated vertebrate axons. Some say that if we did not form myelin, our brains would have to be ten times larger and we would need to consume ten times more food to maintain a nervous system comparable to the one we have. And probably one reason that invertebrates have not developed more complex nervous systems is that their glial cells do not form myelin around axons—they are limited in the number of fast-conducting axons in their nervous systems.

In squid, huge giant axons have evolved to mediate escape responses. (Figure 5.1) The giant axons in squid may be as large as 1 millimeter in diameter. Indeed, the axons are so large that they were once thought to be blood vessels running in squid nerve axon bundles. In the mid-1930s, an English scientist, John Zachary ("JZ") Young, who was spending the summer at the MBL, realized that the long giant axon running down the length of the animal contained no blood cells and that it appeared similar histologically to the smaller axons surrounding it. He proposed that it was a nerve, and with physiologists at the MBL, including H. K. Hartline, he proved that it was indeed a giant nerve cell axon. When this giant axon generates action potentials, powerful muscles throughout the squid's body contract and squirt water from one end of the animal, enabling it to escape by jet propulsion. These muscles have to activate quickly—hence the giant axons.

Once the giant axon was identified as a nerve, it underwent intense investigation. In the late 1930s, scientists showed that by squeezing out the contents of a giant axon, the ionic and other constituents inside cells could be determined: they showed that K^+ levels are high inside the axon and Na^+ and Cl^- concentrations are low. Next physiologists inserted electrodes inside the axon and measured the voltage across the membrane when it was at rest—the resting potential—and when an action potential was generated. These measurements clarified that K^+ is the key ion for establishing the resting potential and Na^+ the key ion for action potential generation. Soon it became possible to free the nerve from the surrounding axons and experiment on an isolated axon, as well as to perfuse the axon with artificial solutions and manipulate inside concentrations of ions or other constituents.

Much of this work was done by two English scientists, Alan Hodgkin and Andrew Huxley, at the Marine Biological Station in Plymouth, England, in the late 1940s and early 1950s. Hodgkin first experimented on the squid giant axon at the MBL in Woods Hole in the summer of 1938, working there with Kenneth ("KC") Cole, an American scientist who pioneered physiological techniques for analyzing the giant axon's response. In 1952, Hodgkin and Huxley provided an analysis of action and resting potentials of the squid giant axon that holds to this day and is applicable to the nerve cell axons of all animals, including humans. For this work Hodgkin and Huxley were awarded the Nobel Prize in 1963.

The squid giant axon remains the largest single adult cell known, so it continues to be of great experimental value. Each summer, scientists flock to Woods Hole or other marine stations around the world to experiment on it.

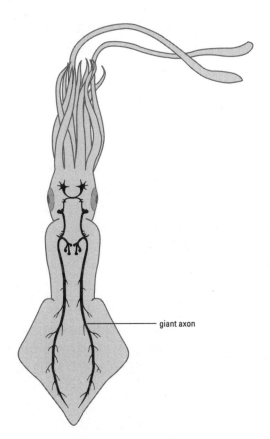

giant axon

FIGURE 5.1 *The giant axon system of the squid. Three pairs of giant axons exist: the brain directly innervates the first giant axons (top), which in turn activate the second giant axons (middle). The second giant axons synapse on the third giant axons, which run the length of the squid's body. The third-order giant axons, formed by the fusion of axons from a number of neurons, are the ones used for most physiological experiments.*

Now the questions have become how the voltage changes across the axonal membrane open Na⁺ channels, and what mechanisms transport substances down the inside of an axon. If substances move down axons only by passive diffusion, it would take about fifteen years to traverse the 1-meter-long axon of the spinal cord motor neuron described in Chapter 1! Axons therefore have devised special transport mechanisms that speed substances along their interiors. This process of *axonal transport* is of particular importance because not all proteins can be made in axons or axon terminals of neurons. Axonal structures depend on proteins made in the neuron's cell body for their maintenance. If axonal transport is disrupted, axonal and terminal function soon fails. Thus, the way axonal transport works is of considerable interest, and we can visualize it operating in squid axons by using video-enhanced microscopy techniques, a technique pioneered at the MBL.

Mach Bands and the Horseshoe Crab Eye

Another exceptionally useful marine invertebrate "discovered" at the MBL is the horseshoe crab, *Limulus polyphemus*. *Limulus* is technically not a crab but a member of the spider family. Like many other creatures in the arthropod (insect) phylum, the horseshoe crab has two prominent faceted eyes. The facets consist of photoreceptive units called *ommatidia*, and each eye in *Limulus* has about twelve hundred of these units. Each ommatidium is made up of about fifteen cells sensitive to light (photoreceptor cells) and one second-order neuron, called an *eccentric cell*, which collects information from the photoreceptor cells. The photoreceptor cells are linked to the eccentric cell by electrical synapses.

When a photon of light is caught by one of the photoreceptor cells, channels open in the cell's membrane that allow Na⁺ to enter, thus depolarizing the cell. The response is the receptor potential, which, via the electrical synapse, passes directly into the eccentric cells. Extending from the eccentric cell is an axon that, if sufficiently depolarized, will fire action potentials, which then travel down the axon. The optic nerve of *Limulus*, extending from the eye to brain, consists primarily of eccentric cell axons. A drawing of a horseshoe crab and its faceted eye and a diagram of a longitudinal section through an ommatidium are shown in

Figure 5.2 (Interestingly, invertebrate photoreceptor cells typically depolarize in response to light, whereas vertebrate photoreceptors hyperpolarize as described in Chapter 4.)

How many action potentials are generated by an eccentric cell axon

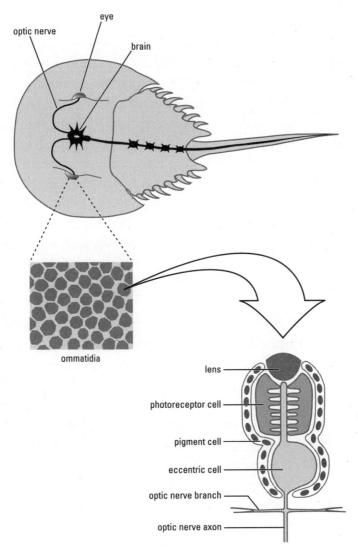

FIGURE 5.2 *The horseshoe crab,* Limulus polyphemus, *and its eye. The optic nerve running from the eye to the brain is located just under the shell, making it highly accessible. The eye is made up of individual units, called ommatidia, that consist of a lens, photoreceptor cells, pigment cells, and an eccentric cell from which the optic nerve axon arises.*

reflects, to a first approximation, the number of light quanta caught by the photoreceptors. But the real-life situation is more complicated, and this discovery led to an elegant explanation of a psychological phenomenon, *Mach bands*, seen at the border between two areas illuminated by different light intensities.

Why was *Limulus* chosen for visual studies? As a college student, H. Keffer Hartline had become interested in visually guided behavior when he found that pill bugs avoid light. His undergraduate research project analyzed this behavior, and in the late 1920s he spent a summer at Woods Hole examining several invertebrates in search of a visual system from which he could record the activity of single axons coming from the eye. What attracted him to the horseshoe crab was its long optic nerve, which runs just underneath the shell from the eye to the anteriorly situated brain (see Figure 5.2). Another attraction is that it is relatively easy to dissect the nerve. In most nerve bundles, connective tissue binds axons together so tightly that it is difficult to tease apart individual fibers. Not so in *Limulus*. Isolating single optic nerve fibers by dissection is straight-forward, as generations of students can attest (see Figure 5.3). Recording from the optic nerve of *Limulus* has long been a favorite laboratory exercise in biology and neurobiology courses.

Hartline's recordings were the first single-cell recordings made from a visual system. He first examined the responses of individual axons to light stimuli and showed that the response paralleled many features of the human visual system. For example, if he used short flashes (less than 1 second), the response he recorded depended strictly on how many photons impinged on the eye. Thus, reciprocal changes in intensity and duration elicited the same response in an axon as long as the flash had the same number of photons. This phenomenon, known as Bloch's law, had long been known in humans; a short, intense flash looks identical to a longer, dimmer flash if both flashes have the same number of photons. These data, then, suggested that the horseshoe crab eye could elucidate aspects of the human eye's function.

When Hartline began his work in the 1930s, and for many years thereafter, he thought that the ommatidia were completely independent units—that they did not interact. One day, as he was recording from an optic nerve fiber and stimulating only the ommatidium where the optic nerve fiber came from, he noticed that when the room lights were turned

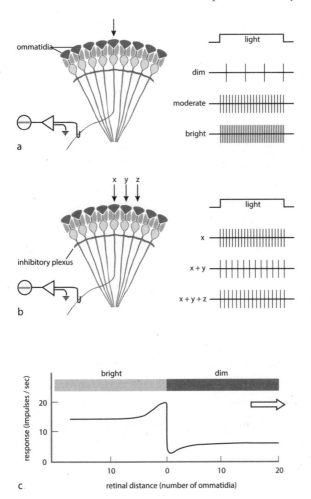

FIGURE 5.3 *Recording from optic nerve axons in Limulus. (a) Illuminating a single ommatidium (X) and recording from the optic nerve axon coming from that ommatidium result in activity levels that depend strictly on light intensity. Dim light evokes a weak response; bright light, a vigorous response. (b) Recording from a single optic nerve axon but illuminating more than one ommatidium results in activity whose level depends not only on light intensity but also on lateral inhibitory effects mediated by the inhibitory plexis. Illuminating X + Y simultaneously reduces the activity generated when only X is illuminated because of the lateral inhibition exerted by Y on X. Illuminating Z in addition to X and Y causes inhibition of Y, partially relieving the inhibition of Y on X. Hence, activity in the optic nerve coming from X increases. (c) The interplay of excitatory and inhibitory effects accentuates contrast at a border between bright and dim lights. Activity is greater along the bright side of the border and lower along the dim side than is activity some distance away from the border.*

on, the recorded axon decreased its activity. Since more light was falling on the eye when the room lights were on, this observation was puzzling. Adding more light to the eye should enhance activity, not dampen it. One explanation was that surrounding ommatidia, now illuminated by the light, inhibited the ommatidium whose activity was being recorded, and this turned out to be true. Hartline and his colleague Floyd Ratliff proved that illuminating surrounding ommatidia could reduce activity in the recorded ommatidium. They called this *lateral inhibition*, and we now know that it occurs in all visual systems.

What purpose does this phenomenon serve for visual processing? Hartline and Ratliff's extensive analyses of lateral inhibition in the horseshoe crab eye in the late 1950s and 1960s demonstrated this elegantly. The inhibitory effects are mediated by fine branches of the eccentric cell axons that arise shortly after the axons emerge from the ommatidia (Figure 5.3). The branches extend laterally to form inhibitory synapses on adjacent eccentric cell axons. The inhibition is reciprocal; an axon inhibits its neighbor, which in turn is inhibited by that neighbor.

The effects of excitation and lateral inhibition on the response of an eccentric cell axon are depicted in Figure 5.3. The axon coming from ommatidium X is recorded with a simple wire hook electrode. If ommatidium X alone is illuminated, the number of action potentials generated per unit time depends on the intensity of light falling on the ommatidium (Figure 5.3a). A dim light evokes few action potentials; a bright light, many more. On the other hand, if the light intensity on ommatidium X is kept constant, illuminating ommatidium Y decreases the firing frequency of the axon from ommatidium X (Figure 5.3b). This is the effect of lateral inhibition. The strength of lateral inhibition between eccentric cell axons depends on how activated the interacting units are and the distance between them; the stronger the illumination, the stronger the inhibition exerted—and near neighbors affect one another more than do distant ones. A further consequence of such a laterally inhibiting network is demonstrated when a third ommatidium, Z, is illuminated. Illuminating Z inhibits Y, which decreases the inhibition of X by Y—a phenomenon called *disinhibition*. What significance does this system of lateral inhibition, reciprocal inhibition, and disinhibition have for vision? Hartline and Ratliff proved convincingly that such interactions can enhance contrast at an edge or border (Figure 5.3c). If a step of light is projected onto an array of ommatidia, the axons coming from the eccentric cells are more active along the bright side of the bright-dim edge than are axons away from the edge; conversely, axons adjacent to the dim side of the edge are less active than axons farther away. This happens because an eccentric cell axon in bright light that is adjacent to a dark border is inhibited weakly by its neighbors in the dimmer light and, conversely, axons along the dimmer side are inhibited strongly by neighbors over the border in bright light. In other words, axons adjacent to a border are inhibited less or more strongly than are axons away from the border. Because of this interplay between excitatory and inhibitory effects, the differences in activity for axons adjacent to the bright-dim border are greater than they are for axons away from the border. Lateral inhibition shapes the signals coming from the axons in such fashion that the intensity differences across the border are enhanced.

A similar enhancement of borders in the human visual system was recognized a century ago by Ernst Mach, an Austrian physicist and psychologist. Called the Mach band phenomenon, it consists of light steps of increasing intensity from left to right. Although each step is even in intensity from one edge to the next, the steps appear lighter along the border adjacent to the darker steps and darker along the borders adjacent to the lighter steps (Figure 5.4). In the human and other vertebrate eyes, Mach band phenomena result from lateral and reciprocal interactions

FIGURE 5.4 *The Mach band phenomenon. Although the reflected intensity of each square is constant from one border to the next, it appears as if each one is lighter along the border with a darker square and darker along the border of a lighter square. The Mach band phenomenon also accentuates the intensity differences between the steps of the series. If the border between two steps is obscured by a pencil or other thin object, the adjacent steps will appear much more similar in intensity than is the case when the same steps meet at a distinct border.*

that occur distally in the retina. The anatomy of the vertebrate eye is different from that of the *Limulus* eye, but the same basic physiological phenomenon is observed; distant photoreceptors inhibit the output of central photoreceptors. In the vertebrate retina, the lateral inhibition is mediated by a separate cell, the horizontal cell.

Learning, Memory, and a Sea Snail

A more recent example of an invertebrate that has provided powerful insights into neurobiological mechanisms comes from the experiments of Eric Kandel, James Schwartz, and coworkers, first at New York University and later at Columbia University's College of Physicians and Surgeons in the 1960s and 1970s. In research carried out on the sea snail *Aplysia californica*, they analyzed two elementary forms of learning and memory: habituation and sensitization. Their findings have had a profound impact on our thinking about mechanisms underlying learning and memory.

Aplysia is a rather unattractive animal; it is about 10 inches long and abounds in tide pools along warm-water coasts. It is a soft animal, not enclosed in a shell, that floats near the water surface and eats seaweed. When disturbed, it secretes jets of dark purple ink.

Why were neurobiologists drawn to this marine organism? *Aplysia* presents several important features. First of all, many cells in *Aplysia* are large—0.8–1 millimeters in diameter—so recording from many of the neurons is relatively easy. The discovery of these large nerve cells in *Aplysia* dates back a century to the early invertebrate histologists. Furthermore, in *Aplysia*, as in many other invertebrates, the nervous system is distributed along the animal rather than collected together in the head. These animals have paired *ganglia*, collections of nerve cells that control one or another behavior. One ganglion might control feeding, another escape behavior, another swimming, and so forth. A ganglion has relatively few neurons, perhaps one thousand to two thousand, so it is possible to record from many of the cells in a single ganglion, identify the cell's function, and explore how the cells are connected together.

When physiologists first began to record from such ganglia in invertebrates, they also found that cells carrying out the same function are located in the same place in the ganglia in animal after animal. Thus, experimenters can construct maps that identify neurons in various ganglia, which greatly facilitates the research. If you want to study a particular cell type, reliable data and diagrams will tell you which ganglion to go to and where in the ganglion the cell will be. Kandel and his colleagues

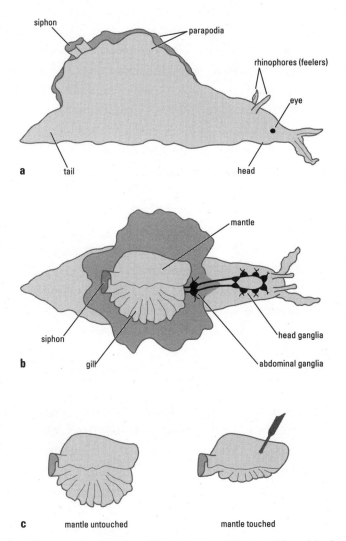

FIGURE 5.5 *The sea slug Aplysia californica, in a side view (a) and looking down on the animal from above when the parapodia are spread apart (b). The gill, lying underneath the mantle, contracts when the mantle or siphon is touched (c). The gill-withdrawal reflex is controlled by neurons in the abdominal ganglia whose approximate location is shown in b.*

took advantage of this to explore the behavior of, and the neural circuitry underlying, the gill-withdrawal reflex.

Figure 5.5 shows *Aplysia* from the side and looking down on the animal when the fleshy coverings on its back side, called parapodia, are separated to reveal the gill. Over the gill's base is the mantle, a protective covering, and associated with the mantle is the siphon, a fleshy extension of the mantle. When the mantle or siphon is touched, the gill rapidly withdraws under the mantle to protect itself. Within a few seconds the gill reappears, but if the mantle or siphon is touched again, the gill disappears again. If the stimuli are continued, the gill continues to withdraw, but with repeated stimuli the gill withdraws less and less. After only four consecutive stimulations spaced one to three minutes apart, gill withdrawal is only about half as extensive as it was originally. And if the stimuli are presented ten times consecutively, at intervals of 30 seconds to a few minutes, gill withdrawal is less than 30 percent of what it was at the first stimulus. This decrease in strength after repeated stimulations is called *habituation*.

It is possible to quantify the extent of gill withdrawal simply by placing a photocell under the gill and illuminating the experimental preparation from above. As the gill withdraws, more of the photocell is exposed, which in turn heightens its output. Figure 5.6 shows records of habituation of the gill-withdrawal reflex. If the mantle receives ten stimulations and then no others, the response will recover in about two hours.

A second modification of the gill-withdrawal reflex is *sensitization*, which happens when a strong stimulus, such as a sharp pinprick, is applied to the head or tail. In sensitization, gill withdrawal is immediately faster, stronger, and longer than in a nonsensitized animal. Recovery after sensitization of the reflex takes minutes to hours, depending on the extent of sensitization.

An interesting feature of habituation and sensitization in *Aplysia* is that they can be induced to last for weeks if a regimen of training stimuli— say, ten consecutive habituating or sensitizing stimuli per day—is given each day for 4–5 days. Here, either habituation or sensitization can be induced to last for a month or so. These phenomena, then, represent long-lasting changes in a neurally mediated response and thereby provide a model for elementary learning and memory. Furthermore, it is of interest that habituation and sensitization may be either brief (lasting hours) or long (lasting weeks) because of the abundant evidence that we and many other animals have two forms of memory: short term and long term. Immediately after we experience something, it is remembered, but these initial memories are quite fragile. A blow to the head or simply being distracted may make a person forget a recent event or recently learned material such as a phone number. Individuals in automobile accidents often fail to remember the accident or what happened 10–15 minutes before it. The survivor of the automobile crash that took the life of Princess Diana showed such an amnesia. He did not remember the crash or events immediately preceding it, but he did remember events of earlier in the evening. Only after a little time (15 minutes or so) do memories begin to become stable and more resistant to erasure. One notion is that short-term memories reflect ongoing neural activity whereas long-term

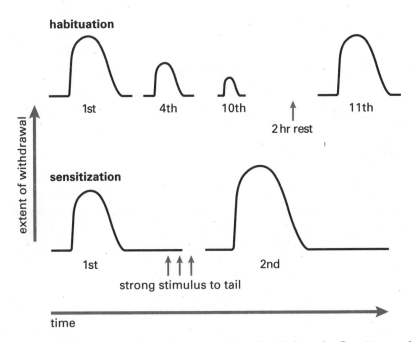

FIGURE 5.6 *Habituation and sensitization of the gill-withdrawal reflex. Repeated touching of the mantle causes rapid habituation of the withdrawal response; the extent of gill withdrawal decreases significantly. Recovery may take 2 hours or more (top). A strong stimulus applied to the tail significantly increases the extent of gill withdrawal; it rapidly sensitizes the response (bottom).*

memories reflect structural changes in the brain—the formation of new synapses, or the structural alteration of existing synapses.

The gill-withdrawal reflex in *Aplysia* is mediated by one ganglion, the abdominal ganglion, which has about fifteen hundred neurons. By recording sequentially from neuron after neuron, Kandel and his colleagues identified thirty-three neurons in the ganglia involved in the basic reflex. Twenty-four are sensory neurons; they are stimulated when the mantle or siphon is touched. Six are motor neurons whose activation prompts the gill to withdraw. The motor neurons drive the muscles that affect gill withdrawal. Three are *interneurons*, whose branches are confined to the ganglion. These neurons receive synaptic input from the sensory neurons and synapse on the motor neurons. In addition, the sensory neurons synapse directly on the motor neurons.

Because the reflex is mediated entirely within the abdominal ganglion, which is just anterior to the gill but on the animal's ventral side, it is possible to remove the ganglion from the animal and study the reflex's neural circuitry when the ganglion is maintained in a small dish. The sensory neurons can be activated by electrical pulses to axons that enter the ganglion; the motor neurons' output can be monitored by recording from axons leaving the ganglion. Furthermore, the isolated ganglion demonstrates habituation. If incoming sensory axons are repeatedly stimulated, the output of the motor neuron axon subsides, just as the gill's withdrawal lessens when the mantle or siphon is repeatedly touched. Figure 5.7 shows a simplified version of the circuitry and the experimental arrangement.

What, then, causes habituation? Because knocking out the interneurons that synapse on the motor neurons had little effect on habituation, Kandel and colleagues focused on the synapse between the sensory neurons and motor neurons. They discovered that this is an unusual synapse in that it releases less and less neurotransmitter when the sensory neurons are repeatedly stimulated. The decrease in motor neuron output can be accounted for by alterations of the sensory neuron synapses on the motor neurons. And this alteration is a diminished release of transmitter following repeated stimulation of the sensory neurons. This was a remarkable finding, that a behavioral modification in an animal could be localized to a single set of synapses! What causes the diminished release of neurotransmitter from the sensory terminals? It is still not

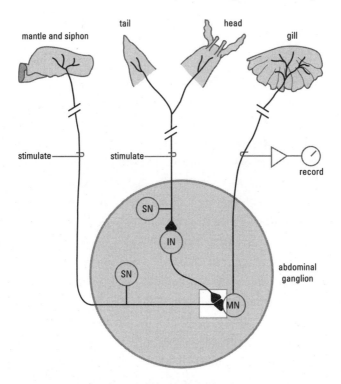

FIGURE 5.7 *A simplified version of the neural circuitry involved in habituation and sensitization of the gill-withdrawal reflex. Sensory neurons (SN) from the mantle and siphon synapse on motor neurons (MN) that innervate the gill. Sensory neurons from the tail or head synapse on interneurons (IN) that make synapses upon the terminals of the sensory neurons coming from the mantle or siphon (box). The interneurons that are activated by the sensory neurons coming from the mantle and siphon and that synapse directly on the motor neurons are not shown because they play no significant role in habituation.*

entirely clear, but Ca^{2+} ions that are responsible for transmitter release (see Chapter 2) are likely involved.

Sensitization also can be induced in isolated abdominal ganglia. Analysis of sensitization has revealed similar but clearer results. Understanding how isolated ganglia are sensitized first requires knowing how inputs from the head and tail interact with the gill-withdrawal reflex circuitry. Sensory input from the head and tail impinges on interneurons in the abdominal ganglion that synapse on the sensory neuron terminals. These interneurons make what are called *presynaptic synapses* in which one set of terminals synapse on another set of synaptic terminals. The circuitry is shown in Figure 5.7.

With this knowledge, experimenters stimulated the sensory axons from the head or tail, which increased neurotransmitter release from the sensory neuron terminals. Hence, reflex alterations were again attributed to a change in one set of synapses—indeed, the same set of synapses—and to a change in the amount of neurotransmitter released, but in this case there was an increase of neurotransmitter release, not a decrease.

How this happens was worked out in exquisite detail, which can be summed up briefly (Figure 5.8). The terminals of the interneurons impinging on the sensory neuron terminals were found to release serotonin, which acts on sensory neuron terminals in classic neuromodulatory fashion: released serotonin activates receptors in the terminal membrane that are linked by a G-protein to the enzyme adenylate cyclase. This causes activation of the enzyme, and the resultant second messenger, cAMP, activates protein kinase A, which phosphorylates ion (K^+) channels in the terminal membrane.

Phosphorylation of this channel keeps the membrane potential of the terminal more positive (depolarized) for a longer time after an action potential comes down the sensory neuron axon. This happens because phosphorylation of the K^+ channels decreases their ability to allow K^+ ions to cross the membrane. Thus, repolarization of the membrane, caused by K^+ leaving the terminal, is slowed down after an action potential comes down the axon (see Chapter 2). With a prolonged depolarization, more Ca^{2+} can enter the terminal and promote neurotransmitter release. Here reflex alteration can be understood down to molecular and ionic levels!

What I've described so far relates to short-term habituation and sensitization. Both can be accounted for by alterations in transmitter release from the sensory neuron terminals. What about long-term habituation and sensitization? Are other factors involved here? Anatomical studies were initially key to learning about these long-term changes. The sensory neurons in the ganglion were injected with a substance that filled the cell and allowed investigators to visualize the sensory neuron axons and their terminals in the electron microscope. The big discovery was that sensory terminals vary among control, long-term-habituated, and long-term-sensitized animals. Terminals were fewer in long-term-habituated animals, and there were more in long-term-sensitized animals; hence, fewer synapses exist between sensory and motor neurons in long-term habituation and more in long-term sensitization. Alterations in individual synapses were noted as well, including changes in their size and the number of vesicles per synapse. These observations indicate that long-term habituation and sensitization bring about structural changes in neurons

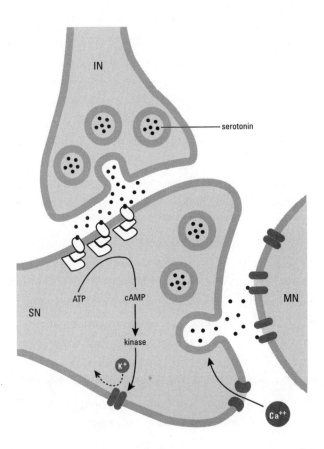

FIGURE 5.8 *A summary scheme of the synaptic interaction giving rise to sensitization. Serotonin released from the interneuron terminal (IN) activates receptors in the sensory neuron terminal (SN) linked through a G-protein to adenylate cyclase, an enzyme that changes ATP to cAMP. The kinase activated by cAMP phosphorylates K+ channels in the sensory neuron terminal membrane, thereby closing them. This results in prolongation of terminal activation and increased release of neurotransmitter from the sensory neuron terminal because more vesicles can bind to the membrane. With more neurotransmitter released, the motor neuron (MN) response is larger and gill withdrawal more extensive.*

and synapses. This implies that new protein synthesis or degradation happens in the neurons because of these processes, and that genes are turned on or off. The evidence is compelling that kinases within nerve and other cells alter protein synthesis and degradation and gene activation and inactivation.

The experiments on *Aplysia* thus provide experimental support for the older ideas that short-term memories reflect alterations in ongoing neural processes and that long-term memories reflect alterations in neural structures. The *Aplysia* model explains elegantly what such changes might entail and how they can come about. In humans, a region of the brain called the *hippocampus* is essential for consolidating short-term memories. Destruction of the hippocampi (we have two, one on each side of the brain) results in individuals who cannot remember things for more than a few minutes (see Chapter 11). From the many studies of hippocampal mechanisms, investigators are finding that similar mechanisms operate in the hippocampus as in the abdominal ganglia of *Aplysia*, when stimuli to the hippocampus result in long-term physiological changes.

Circadian Rhythms, Behavioral Genetics, and Flies

Many biological processes, including behaviors, are regulated by an internal clock, or *circadian rhythm*. *Circ* is Latin for "around" and *dian* for "day." Sleep is a familiar example. Rats, being nocturnal, sleep during the day and are active at night. If their circadian rhythms are disrupted, rats tend to sleep for short periods throughout both the day and night. The importance of a circadian clock in human sleeping behavior is illustrated dramatically when we travel by air over several time zones. Particularly, if we go places halfway around the world, we will be awake much of the night and sleepy during the day. We gradually adjust to the new time zone, but the so-called jet-lag time can be unpleasant. And we go through the same process when we return home.

The initial insights on how such rhythms are generated came from studies on fruit flies by Seymour Benzer and his colleagues at the California Institute of Technology beginning in the 1970s. Benzer was a geneticist who mutagenized the fruit fly, *Drosophila melanogaster*, by a chemical that alters single genes. Whereas typical circadian rhythms oscillate over a period of 24 hours, he and his colleagues found some fruit-fly mutants whose cycle was longer than 24 hours, others that were much shorter, and still others that had no rhythms at all; they were arrhythmic. It turned out that mutations in one gene underlay all these phenotypes. The gene involved, termed *period* (or *per*), codes for a novel protein. Particularly noteworthy is that the synthesis and accumulation of both *per* mRNA

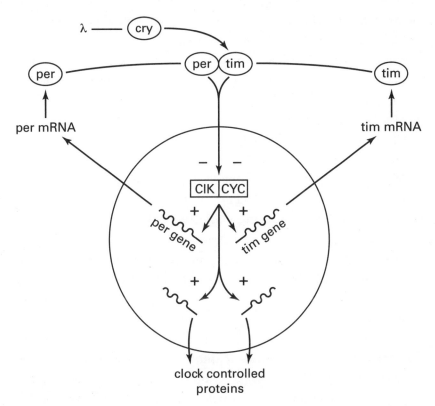

FIGURE 5.9 *Schematic drawing of feedback loops underlying circadian rhythm generation in Drosophila. The transcription factors clk and cyc activate the promoter regions of the per and tim genes, which results in the accumulation of* per *and* tim *mRNA and protein in the cytoplasm. When protein levels are sufficiently high,* per *and* tim *proteins dimerize and the complex translocates to the nucleus, where it inhibits the activity of the clk-cyc protein dimer complex, thereby reducing levels of* per *and* tim *mRNA and protein. The clk-cyc complex also interacts with the promoter regions of other genes that code for clock-controlled proteins. Finally, light activates cry, which promotes the turnover of* tim *and* per, *resulting in resetting of the clock.*

and the *per* protein vary in the short- and long-day mutants, as does their circadian behavior; synthesis and accumulation of *per* mRNA and protein are faster in short-period mutants and slower in long-period mutants.

What is going on is that the *per* protein controls expression of the *per* gene by a negative feedback loop. The arrhythmic mutants produce a defective protein that leaves the *per* gene permanently on. The key factors in generating the rhythm are appropriate time constants for synthesizing and degrading mRNA and protein and a sufficient buildup of protein before inhibitory feedback occurs. In normal animals all of the requirements are met, but not in the mutant animals.

Subsequent work has shown that the cellular mechanisms underlying circadian rhythms are more complicated than the simple picture described above, although the early ideas are basically correct. The diagram in Figure 5.9 incorporates many of the more recent findings, but even this figure is somewhat simplified. It turns out that for the *per* protein to inhibit expression of the *per* gene, it must first dimerize with another protein, called *tim* for "timeless." This complex then translocates into the nucleus of the cell, where it depresses the activity of another dimer consisting of two proteins, clock (*clk*) and cycle (*cyc*). The *clk-cyc* dimer interacts with the promoter regions of both the *per* and *tim* genes, activating them and leading to the expression and accumulation of *per* and *tim* mRNA and protein in the cytoplasm. When sufficient *per* and *tim* proteins accumulate, they dimerize, permitting their translocation into the nucleus and inhibition of *clk-cyc* activity. This leads to decreased synthesis of *per* and *tim* proteins and eventually decreased levels of the *per-tim* dimer. Inhibition of the *clk-cyc* dimer is then relieved and renewed activation of the *per* and *tim* genes ensues. The slow accumulation of the *per* and *tim* proteins appears to account for much of the 24-hour length of the circadian clock.

Two other features of the circadian clock mechanism are illustrated in Figure 5.9. First, the *clk-cyc* dimer can also activate other genes, thus explaining the circadian regulation of a variety of clock-controlled proteins. Second, light can reset the clock through a protein called cryptochrome (*cry*), which promotes turnover of the *tim* protein. The *tim* protein stabilizes the *per* protein. When *tim* levels decline, so do *per* levels, and the clock is reset. Genes homologous to *per*, *tim*, *clk*, and *cyc* have now been found in a number of species, including mammals. The present view is that mechanisms similar to those in *Drosophila* probably underlie circadian rhythmic activity generally in nature. A Nobel Prize was awarded in 2017 to disciples of Benzer who worked out the details of the circadian clock mechanisms; sadly, Benzer, who pioneered the field of behavior genetics in Drosophila, passed away in 2007.

The foregoing might suggest that only invertebrates have given us powerful insights into neural mechanisms underlying behaviors. This is far from the truth. Indeed, we have learned much from studying vertebrates, and neuroscientists now routinely investigate neural mechanisms and behaviors in mammals such as mice and monkeys. Again, though, neuroscientists have often turned to simpler vertebrates, such as fish or frogs, to investigate neuronal mechanisms or well-defined and perhaps simpler parts of the nervous system. My own research is on the vertebrate retina, a part of the brain displaced into the eye during development of the embryo. Its organization and anatomy are quite well understood. Furthermore, certain animals such as the mudpuppy, a salamander-like animal, have large cells for facilitating recordings from the neurons, as is the case in *Aplysia*. Further, zebrafish can be readily mutagenized as Benzer did with fruit flies, and some of these mutants have provided insights regarding certain vertebrate behaviors. The point, then, is that neuroscientists take advantage of the enormous variety of animals that have evolved on this planet in their efforts to understand the brain. Neurobiological mechanisms are similar in all species, which makes this approach useful. Ultimately, though, it is detailed knowledge of the human brain that we seek.

6

Vertebrate Brains

....

Christina was a strapping young woman of twenty-seven, given to hockey and riding, self-assured, robust, in body and mind. She had two young children, and worked as a computer programmer. . . . She had scarcely known a day's illness. Somewhat to her surprise, after an attack of abdominal pain, she was found to have gallstones, and removal of the gallbladder was advised.

She was admitted to hospital three days before the operation date, and placed on antibiotics for microbial prophylaxis.

The day before surgery Christina, not usually given to fancies or dreams, had a disturbing dream of peculiar intensity. She was swaying wildly in her dream, very unsteady on her feet, could hardly feel the ground beneath her, could hardly feel anything in her hands, found them flailing to and fro, kept dropping whatever she picked up.

She was distressed by this dream, so distressed that we requested an opinion from the psychiatrist. "Pre-operative anxiety," he said.

But later that day *the dream came true*. Christina did find herself very unsteady on her feet, with awkward flailing movements, and dropping things from her hands.

But the day of surgery Christina was still worse. Standing was impossible—unless she looked down at her feet. She could hold nothing in her hands, and they "wandered"—unless she kept an eye on them. When she reached out for something, or tried to feed herself, her hands would miss, or overshoot wildly, as if some essential control or coordination was gone.

She could scarcely even sit up—her body "gave way." Her face was oddly expressionless and slack, her jaw fell open, even her vocal posture was gone.

"Something awful's happened," she mouthed, in a ghostly flat voice. "I can't feel my body. I feel weird—disembodied."

—Excerpted from Oliver Sacks, *The Man Who Mistook His Wife for a Hat* (New York, NY: Harper and Row, 1970)

Christina had lost all position sense—sensory information coming from the muscles, joints, and tendons telling the brain the status of her limbs and trunk, including their positions. Such information is called *proprioceptive* and is largely unconscious. We are unaware of most of this sensory input from the limbs and body, but it is essential for the brain to "sense" the body. As Christina noted, "I lose my arms. I think they are in one place and I find they're [in] another."

A quite peculiar and selective inflammation of the proprioceptive axons had occurred all along Christina's spinal cord and brain. Over the next few days the inflammation gradually subsided, but the axons did not recover. Christina had permanently lost virtually all proprioception. However, Christina did gradually manage to deal with her deficit, by using her eyes and to some extent her ears to monitor her movements, to keep track of her limbs, and to modulate her speech. But it was an excruciatingly slow process. It took nearly a year before she could leave the hospital, rejoin her family, and resume her computer work. She still was by no means normal but had learned to compensate. How did she feel? It was difficult for her to describe her state except in terms of her other senses. "I feel my body is blind and deaf to itself . . . it has no sense of itself."

In vertebrates, the nervous system is divided into two parts: the central nervous system, made up of the spinal cord and the brain proper, and the peripheral nervous system, consisting of all the nerves and nerve cells that lie outside of the spinal cord and brain. The peripheral nervous system is largely made up of motor axons and sensory neurons. The axons of the motor neurons carry the instructions from the central nervous system out to the muscles of the head, face, limbs, and body, allowing us to make purposeful movements and to act. Sensory neurons, carrying information into the central nervous system from the various specialized sense organs, inform the brain about the rest of the body and the world beyond. We are not aware of all of the sensory information coming into the brain, including most proprioceptive input. Without proprioceptive sensory input, the brain is isolated from much of what is going on in the body, as happened to Christina.

Although neuronal mechanisms in the central and peripheral nervous systems are virtually identical, there is a surprising difference in the glial cells of these systems in mammals that has important consequences for us. In the peripheral nervous system, the glial cells that surround nerve cell axons and form the myelin sheath are called *Schwann cells*, named after their discoverer, Theodore Schwann. He is best known for his theory, announced in 1839, that all animals are formed of cells. In the central nervous system, the glial cells that form myelin are known as *oligodendrocytes*. Why the difference in glial cells is important relates to the regrowth of axons after they are cut or injured. Whereas axons in the peripheral nervous system will regenerate, axons in the central nervous system of mammals will not, a difference related to the surrounding glial cells.

If a limb or other part of the body served by the peripheral nervous system is severed and then surgically reattached, function and sensation can make a remarkable recovery. This is a slow process requiring many months, but sensory and motor neuron axons eventually regenerate and make appropriate connections. Axons in the central nervous system, in contrast, do not regrow after injury or cutting. This is best exemplified by spinal cord injuries that cut or crush nerve cell axons. Paralysis and loss of sensation occur below the level of injury, and this condition is permanent. Such individuals never can walk again, and if the spinal cord injury is just below the neck, they may also lose the use of their arms. They may also need assistance breathing and require a respirator. The actor Christopher Reeves, whose spinal cord was crushed in a riding accident, provides an example of this tragic injury.

Albert Aguayo and his colleagues in Montreal have demonstrated that if severed central axons are put in contact with Schwann cells, they will regenerate. The experimenters remove a piece of nerve from an animal's leg or another part of its peripheral nervous system, which prompts the axons in the nerve to degenerate. Then the investigators place the remaining Schwann cells around a severed nerve, such as the optic nerve, of the central nervous system. The result is that some axons in the optic nerve will regenerate and reinnervate part of the brain and some sight is partially restored. The importance of such findings for the possibility of treating spinal cord and brain injuries cannot be overstated.

Why do Schwann cells allow for axonal regeneration but oligodendrocytes do not? This is still not well understood. Two theories, not mutually exclusive, are being tested. One proposes that Schwann cells release one or more substances that promote axon regeneration; the other theory states that oligodendrocytes release a factor that forbids the regeneration of axons. Some evidence supports both theories, and both situations may hold. Also, central nervous system axons do regenerate in many cold-blooded vertebrates, so comparing oligodendrocytes from mammalian and nonmammalian species may provide more clues to the glial cell factors involved. The critical and surprising conclusion is that axonal regrowth is regulated by glial cells, not by the neurons themselves.

The Central Nervous System

Figure 6.1 is a drawing of the central nervous system—the spinal cord and brain—of a human viewed from the back side. The spinal cord is housed in the vertebral column; the brain is in the skull. Unlike most invertebrate nervous systems, which are distributed along the animal, the nervous system in vertebrates such as ourselves is highly centralized. In us, all higher nervous system functions, including movement initiation, perception, memory, learning, and consciousness, are carried out within the brain. The spinal cord, in contrast, serves three major roles: simple reflexes, such as the knee-jerk reflex in which a tap just below the knee makes the lower part of the leg kick out, are mediated within the spinal cord; the neural circuitry for rhythmic movements such as walking or

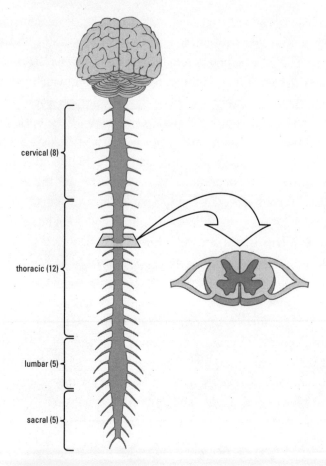

cervical (8)

thoracic (12)

lumbar (5)

sacral (5)

FIGURE 6.1 *The central nervous system consists of the brain and spinal cord. Bundles of axons extend from the spinal cord in pairs—thirty pairs in all—and carry sensory information into the spinal cord or motor information out from the spinal cord to the muscles. Four regions along the spinal cord are recognized— cervical, thoracic, lumbar, and sacral. The number of nerve bundles exiting from each region is indicated in parentheses. A cross section of the spinal cord in the thoracic region is indicated on the right.*

scratching are found in the spinal cord; and the cord itself serves as a conduit for sensory information from the periphery to the brain and for motor information from the brain to the motor neurons found in the spinal cord.

As shown in Figure 6.1, thirty pairs of nerve bundles extend from the spinal cord. These nerves consist of sensory and motor axons. Sensory information from the body comes into the spinal cord by way of axons in these nerves; motor information goes to the muscles via other axons in the nerves. The left member of a pair of nerves extending from the spinal cord serves a portion of the left side of the body, the right member of the pair serves a portion of the right side of the body. Much of the sensory information entering the spinal cord is called *somatosensory*—conveying information from touch, temperature, pressure, and pain receptors in the skin and deep tissues of the limbs and trunk. In addition to the thirty pairs of spinal cord nerves, twelve pairs of nerves enter the brain directly. These *cranial nerves* carry sensory and motor information related to the head. Three cranial nerves carry visual, auditory, or olfactory sensory information; three are devoted to controlling eye movements; and four are a mix of sensory and motor axons innervating the face, tongue, neck, and jaw. Finally, two cranial nerves innervate internal organs such as the heart, lungs, and digestive system.

Spinal Cord

A slice through the spinal cord is depicted in Figure 6.2. In the cord's interior is a butterfly-shaped region that stands out from surrounding regions by its grayish hue (termed *gray matter*). The cell bodies of neurons and their dendrites are in this interior region, and most of the synapses are made here also. The surrounding, whiter region (*white matter*) contains bundles of axons running up and down the cord. The axonal bundles are whitish because of the myelin sheaths covering them.

Sensory information comes into the cord dorsally—from the back side—while motor information exits the cord ventrally, from the front side. The cell bodies of sensory neurons are collected together in ganglia that lie just outside the cord. The sensory neuron axons entering the spinal cord can do any one of three things. The can ascend the cord, carrying sensory information to higher cord levels or to the brain itself; they can synapse on interneurons found within the gray matter; or they can synapse directly on motor neurons found ventrally in the gray matter.

When a sensory neuron axon directly innervates a motor neuron, as in Figure 6.2, a simple reflex circuit is created. The sensory neuron directly activates a motor neuron, and the result is a behavior independent of

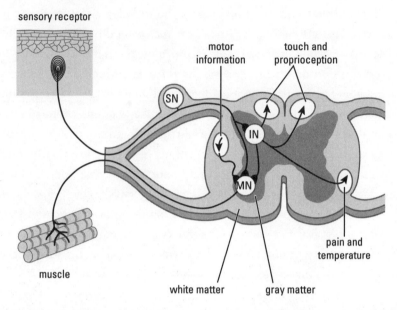

sensory receptor

motor
information

touch and
proprioception

SN

IN

MN

pain and
temperature

muscle

white matter   gray matter

FIGURE 6.2 *A cross section of the spinal cord. Axons run up and down the cord around the periphery of the cord in the white matter. The neurons, dendrites, and synapses are in the central gray matter. Sensory information enters the cord dorsally, and synapses of the sensory neurons (SN) may be made onto interneurons (IN) or directly onto motor neurons (MN). The latter situation is the basis of a simple monosynaptic reflex. Motor neuron axons exit the cord ventrally to innervate muscles. Sensory information travels up the cord via axons of the interneurons. These axons run along specific paths in the white matter depending on the type of sensory information they are carrying. Interneurons may innervate motor neurons, and the motor neurons also receive input from axons descending down the cord (left side of drawing). For simplification, information entering and exiting the cord is shown only for one side, and the interneuron is shown having multiple axons.*

much of the rest of the nervous system. The knee-jerk reflex is such a reflex; a one-synapse reflex is called a *monosynaptic reflex*. Reflexes also involve interneurons and hence are polysynaptic; indeed, even the knee-jerk reflex involves interneurons that inhibit the muscles that are opposite in their action to those activated in the knee-jerk reflex. Reflex circuitry can become quite complicated, involving many interneurons and several levels within the spinal cord.

The axons that run up and down the cord are segregated according to whether they are sensory or motor axons and what type of sensory information they are carrying. Axons carrying pain and temperature information are in one region of the cord, touch in another region, and so forth. Motor axons are also segregated, descending in areas separate from sensory areas, some of which are indicated in Figure 6.2.

A curious feature of the organization of the vertebrate nervous system is that much of the sensory information ascending the spinal cord does so on the opposite side of the cord from where it enters, and it eventually reaches the brain on this opposite side. Hence, information from the body's left side is processed mainly on the brain's right side, and vice versa. The same holds for motor information. The brain's right side initiates movements for the body's left side, and vice versa. Why one side of the brain controls the opposite side of the body is a mystery, and no one has a compelling explanation of why this is so. Its consequences are that an injury on one side of the brain or spinal cord affects sensation or motor control on the body's opposite side: a stroke on the brain's left side paralyzes the body's right side, and vice versa.

The sensory information ascending the spinal cord described so far is all conscious information; we are aware of it. It reaches brain regions that participate in conscious perception. The cord also has ascending sensory information of which we are not aware. This sensory information from our muscles, joints, and tendons is critical for efficient and coordinated movement and essential for proper brain function. It tells the brain the position of our limbs and their status. The degree of muscle contraction is signaled to the brain, for example, but we are not aware of such sensory information; it does not enter our consciousness. This information is called *proprioceptive* after the Latin *proprius*, meaning "one's own." Without it, patients are severely debilitated, as the story of Christina at the beginning of this chapter illustrates.

Proprioceptive information runs up the spinal cord dorsally. In the late stages of syphilis these tracts partially degenerate, and patients suffering from this degeneration typically have exaggerated movements like high-stepping when they walk. Fortunately, with the advent of antibiotics only rare cases of this degeneration are seen today. But even in the 1950s, when I was a medical student, patients with the characteristic gait of this degeneration were not uncommon.

Another degeneration involving the spinal cord that is no longer a problem, at least in the developed world, is poliomyelitis. The polio virus invades motor neurons in the ventral regions of the spinal cord and destroys them, leading to the paralysis of the muscles these neurons innervate. All muscles of the limbs and trunk are controlled by motor neurons in the spinal cord. They are, in the words of Charles Sherrington, an eminent English scientist working early in the last century who was particularly interested in spinal cord reflexes, the "final common pathway" regulating movement. They are acted on by axons descending the spinal cord carrying information from the brain, directly from sensory neurons entering the spinal cord, and by spinal cord interneurons. Their output determines the movement of a part of the body, and their loss results in paralysis of that part.

Finally, within the spinal cord is the circuitry underlying several rhythmic movements such as walking and scratching. The evidence for this comes from experiments on cats, which received a lesion in the lower part of the brain that separates this region and the spinal cord from higher brain regions. The animals can stand, and if placed on a treadmill will walk if a specific group of neurons in their lower brain stem is stimulated. These neurons appear to provide a signal, a command, to interneurons in the spinal cord, which then interact in such a way that they give motor neurons appropriate synaptic input for coordinated walking. We only vaguely understand the circuits underlying these rhythmic motor behaviors, but many investigators are trying to work out their circuitry.

The Brain

Figure 6.3 shows a longitudinal section through the human brain and part of the spinal cord. Essentially this section is a cut down through the middle of the brain, and it reveals several of the major brain structures. As expected, the structure of the brain is complex, consisting of a number of subdivisions that differ strikingly in their anatomy. And such a longitudinal section fails to reveal certain important brain regions.

To sort out the various structures, it is helpful to divide the brain into three regions: *hindbrain, midbrain,* and *forebrain.* The hindbrain emerges from the spinal cord and has three main structures, the *medulla, pons,* and *cerebellum*. The midbrain sits between the hindbrain and forebrain, and in humans it is relatively small. The forebrain is by far the most prominent part of the human brain and is divided into two regions. One includes the *thalamus* and *hypothalamus* and the other includes the *basal ganglia* (not shown in Figure 6.3) and the *cerebral cortex*. The basal ganglia and cerebral cortex are called collectively the *cerebrum*. A prominent band of axons, the *corpus callosum*, lies centrally under the cerebral cortex and carries information between the brain's right and left sides.

The brains of cold-blooded vertebrates such as fishes or frogs provide insights into the evolution of the vertebrate brain. (Figure 6.4)The brains of these animals have a small cerebrum relative to the rest of the brain. In most cold-blooded vertebrates the cerebrum is concerned primarily with the analysis of one sensory modality, smell. In mammals the cerebrum is greatly expanded, and most higher neural functions are centered

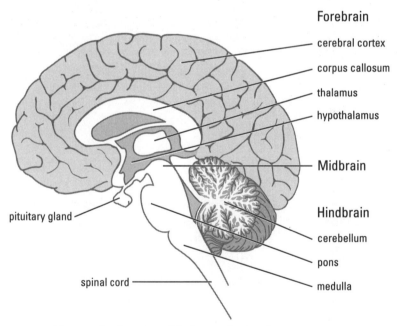

FIGURE 6.3 *A longitudinal section of the human brain, showing its major structures. The hindbrain emerges from the spinal cord and consists of the cerebellum, pons, and medulla. The midbrain encompasses a relatively small part of the primate brain, whereas the forebrain is very prominent. Major structures of the forebrain include the cerebral cortex, corpus callosum, thalamus, and hypothalamus. The pituitary gland is found just below the hypothalamus.*

in the cerebral cortex. The proportionally larger forebrain, especially the cerebral cortex, is by far the most significant difference we observe when comparing higher mammals with cold-blooded vertebrates. The evolution of the vertebrate brain is thus mainly about the forebrain; it has evolved from a structure concerned mainly with analyzing just one sense to, in humans, the seat of sensation, memory, intelligence, and consciousness.

In cold-blooded animals the midbrain is much more prominent, and the brains of many of these animals are dominated by a structure called the *tectum*. The tectum receives visual and other sensory input, and tectal neurons project to the spinal cord, where they synapse onto motor neurons. The tectum is the key structure in nonmammalian species for integrating sensory inputs and for controlling *motor outputs*.

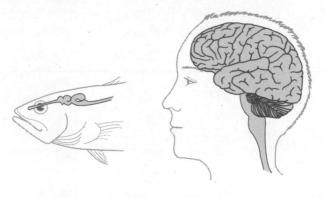

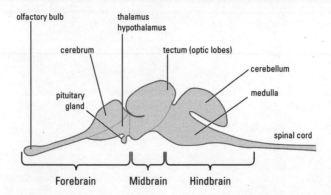

FIGURE 6.4 *A comparison of the brains of fish and man. The human brain is very much larger, but the same structures exist in both brains (compare with Figure 6.3). However, the midbrain in the fish is relatively much larger and is dominated by the tectum.*

In higher mammals, the midbrain and tectum are less important brain structures. Sensory processing, sensory-motor integration, and the initiation of motor activity are in the cerebral cortex. The midbrain and tectum in the mammal mediate noncortical visual reflexes like pupillary and eye movements. In higher mammals the tectum also helps to coordinate head and eye movements.

Medulla and Pons

The medulla links the spinal cord to the brain. Thus, running through the medulla are numerous ascending (sensory) and descending (motor) axon bundles. Half of the cranial nerves enter the brain in the medulla. But within the medulla are critical nuclei concerned with regulating vital body functions, including respiration, heart, and gastrointestinal function. The bullet that killed President Kennedy entered this part of his brain and destroyed many of these vital regulatory centers. Without these critical nuclei, death inevitably ensues. Nuclei that control the head, face, eyes, and tongue are also in the medulla.

In addition to the discrete nuclei in the medulla, clusters of neurons are spread diffusely throughout the medulla, especially ventrally. These neurons make up what is called the *reticular formation*. Some extend their axons widely throughout the brain, and their synaptic terminals typically contain neuromodulatory substances—monoamines or peptides. Reticular formation neurons exert their effects on virtually all parts of the brain. They are involved in arousal and controlling levels of consciousness; lesions in the reticular formation can cause animals, including humans, to lose consciousness or to fall into a stupor from which they cannot be aroused.

The pons (from the Latin word for "bridge") contains neurons that receive input from the cerebral cortex. They relay this information to the opposite side of the cerebellum. The pons thus serves as a switchboard between the cerebral cortex and cerebellum, mediating the crossing of most motor information from one side of the brain to the other. Some reticular formation neurons are also found in the pons.

Hypothalamus

The hypothalamus and the medulla are the brain's principal regulatory centers. Nuclei in the hypothalamus mostly regulate basic drives and acts such as eating, drinking, body temperature, and sexual activity. The hypothalamus also plays a role in emotional behavior, which is discussed in detail in Chapter 12. Lesions in the hypothalamus, or stimulation of parts of it, can lead to irritability or even aggressive behavior. Yet lesions or stimulation of other parts of the hypothalamus can lead to placidity.

The *pituitary gland* sits underneath the hypothalamus, and another task for the hypothalamus is to regulate the pituitary gland's release of hormones. Hypothalamic neurons do this by releasing small peptides that promote or inhibit the release of the pituitary hormones. The pituitary hormones, by way of the bloodstream, regulate hormone release from glands found elsewhere in the body, such as the thyroid and adrenal glands; or they may exert direct effects on tissues. Examples of the latter are growth hormone and oxytocin, the mammary gland milk-releasing hormone. Hypothalamic nuclei, along with nuclei in the medulla, also help to control the autonomic nervous system. This system regulates internal organs, including the heart, digestive system, lungs, bladder, blood vessels, certain glands, and the pupil of the eye; it is discussed in detail in Chapter 12.

Cerebellum

The cerebellum coordinates and integrates motor activity. The command for a skilled movement comes from the cerebral cortex, but the cerebellum must coordinate a motor command with sensory information to ensure smooth movement. Thus, the cerebellum receives input from the cortex via the pons and sensory input from the spinal cord and other sensory systems. Much of the proprioceptive sensory information coming up the spinal cord goes to the cerebellum. The cerebellum compares the various inputs, integrates them, and sends signals to the motor neurons in the spinal cord to accomplish smooth, coordinated movements.

Lesions of the cerebellum typically result in jerky, uncoordinated movements. Movement initiation may be delayed, or movements may be exaggerated or inadequate. The cerebellum is also responsible for the learning and memory of motor tasks, such as riding a bicycle. Thus, cerebellar lesions can also interfere with learning and remembering motor skills.

Thalamus

The thalamus has numerous nuclei whose role is to relay sensory information to the cerebral cortex. Other thalamic nuclei send information to the cortex about motor activity. Individual nuclei thus convey specific

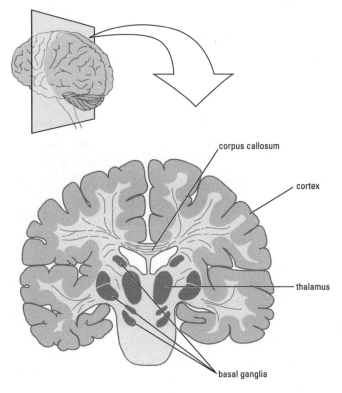

FIGURE 6.5 *A vertical section through the brain cut according to the diagram shown at the top. The cortex is a dense cellular layer, about 2 millimeters thick, covering the two cortical hemispheres. The infoldings of the cortex increase the brain's surface area and therefore the amount of cortex. The corpus callosum is a thick band of axons that connects the two hemispheres. The basal ganglia and thalamus are subcortical structures. (In this drawing, the thickness of the cortex is exaggerated.)*

sensory or motor information. The lateral geniculate nucleus, for example, transmits visual information to the cortex. Most optic nerve axons terminate in this nucleus, and the neurons in the lateral geniculate nucleus then project to the area of the cortex that processes visual information.

The thalamic nuclei receive input back from the cortex and also from the reticular formation in the medulla. These nonsensory inputs control the flow of information from the thalamus to the cortex. Thus, a key role carried out by the thalamus is to gate information flow from the spinal cord and lower brain structures to the cerebral cortex.

Basal Ganglia

Lateral to the thalami on both sides of the brain are five prominent nuclei known as the *basal ganglia*. These are concerned primarily with the initiation and execution of movements. The position of the basal ganglia is shown in Figure 6.5, which represents a vertical section through the middle of the brain. The basal ganglia receive input from the cortex and feed information back to the cortex by way of the thalamus.

Lesions in the basal ganglia cause characteristic movement abnormalities. Tremors and repetitive movements are frequently seen in such patients, and individuals with basal ganglia defects have difficulty in initiating movements. Rigidity of the limbs is another common symptom.

One disease of the basal ganglia is, as we have noted, Parkinson's disease, and another is *Huntington's disease*, which is inherited and is also related to the loss of specific neuroactive substances within certain ganglia. In Parkinson's disease dopaminergic neurons are lost, and in Huntington's disease GABA and acetylcholine are lost; neurons containing these substances characteristically degenerate. Early in Huntington's disease, which typically begins in middle age, patients experience small but uncontrolled movements of the arms, legs, torso, and face. These spontaneous movements gradually worsen, the patients have difficulty swallowing, and they lose their balance and become very unsteady until they are confined to bed. Accompanying the movement disorder are mood swings, depression, irritability, and eventually loss of cognition and dementia. Death generally ensues 15–20 years after onset of the disease.

Approximately 25,000 Americans are affected with Huntington's disease, with another 125,000 individuals at risk for it. The disease is inherited in a dominant fashion; that is, one copy of the defective gene from either the father or mother will cause the disease. Thus, 50 percent of the offspring of an affected individual will inherit the disease. The defective gene that causes the disease has been isolated and its sequence determined. From the gene sequence, the normal protein structure it codes can be predicted, but the predicted protein is unlike any known protein, hence its exact function is a mystery. The protein is widely scattered throughout the nervous system, which suggests that it plays a significant role, but what that role may be is presently unknown.

Cerebral Cortex

Without doubt, the most important part of the brain for us is the cerebral cortex. It is in the cortex that virtually all of our higher mental functions are localized. The initiation of skilled movements, perception, consciousness, memory, and intelligence all depend critically on the cerebral cortex. The cortex is divided into two *hemispheres*, each of which is subdivided into four lobes termed *frontal, parietal, occipital,* and *temporal*. Figure 6.6 shows a surface view of the left hemisphere of the human cerebral cortex.

The cortical neurons are close to the surface, where they form a layer about 2 millimeters (less than 0.1 inch) thick that covers each hemisphere. To increase surface area, and thereby have more cortical neurons, the cortex is extensively infolded in higher mammals and ourselves. How this works is shown in Figure 6.5. In humans, the total area of the cortex is about 1.5 square feet if spread out, and under each square millimeter of cortex are about 100,000 neurons. Thus, the human cortex as a whole has about 10^{10} or 10 billion neurons.

Specific roles are carried out within each cortical lobe. The frontal lobes, for example, are concerned primarily with movement, smell, planning, and programming; the parietal lobes, with somatosensory information processing; the occipital lobes, with vision; and the temporal lobes, with hearing and memory consolidation. But each lobe consists of many subareas, whose roles vary widely. For example, within each of the lobes are areas devoted to the initial processing of one sensory modality or to the initiation of skilled movements. These *primary sensory areas* and the primary motor projection area are shown in Figure 6.6. A deep infolding, called the central *sulcus*, separates the motor projection area from the primary somatosensory area, and also the frontal lobe from the parietal lobe. Another deep infolding, called the lateral sulcus, separates the temporal lobe from the frontal and parietal lobes. Many more infoldings are present in the human cortex than are shown in Figure 6.6.

Along the primary somatosensory cortex, sensory information from parts of the body is received and analyzed. Thus, a rough representation of the body surface exists on the primary somatosensory cortex (Figure 6.7). This representation is, however, not proportional. More cor-

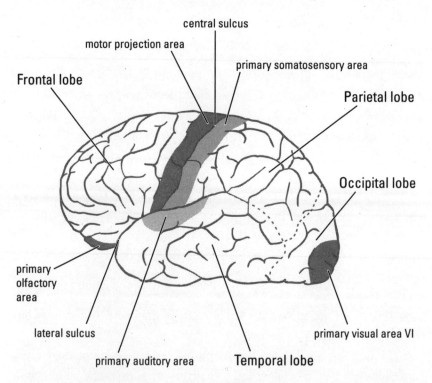

FIGURE 6.6 *A left-side view of the outside surface of the cortex. The four cortical lobes—frontal, parietal, occipital, and temporal—are shown, as are the primary sensory-processing areas and motor projection area. Two of the deep infoldings (sulci) are indicated—the central sulcus separates the frontal and parietal lobes, and the lateral sulcus separates the temporal lobe from the frontal and parietal lobes.*

tical area is devoted to body parts where sensation is more acute, such as the face and hand, and less cortical area is devoted to regions that receive less sensory innervation, such as the back or upper leg. The same situation holds for the primary motor projection area and the primary visual area. The hand and face occupy a disproportionate amount of the primary

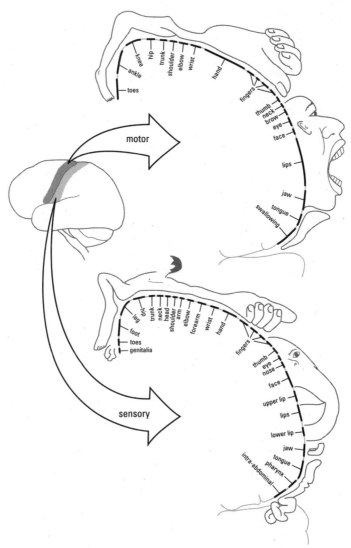

FIGURE 6.7 *The representation of the face and body on the primary motor cortex (top) and primary somatosensory cortex (bottom). Areas such as the hands and face with more innervation or where sensations are more acute, or that have finer movements, have larger cortical representations.*

motor projection area, and the central (foveal) region of the retina, which mediates high-acuity vision, encompasses a large proportion of the primary visual area. One can relate the sensitivity of sensation or the degree of motor control to the amount of cortex it occupies or, simply, the number of neurons involved.

Away from the primary sensory or motor projection areas are secondary or *association areas* that participate in higher levels of processing or the processing of one aspect of a sensory stimulus. As many as thirty areas may be involved in the processing of visual information. Damage to the primary visual area, area V1, causes the loss of all visual perception. That is, a patient is not aware of seeing anything in the part of the visual field that projects to the damaged cortical area. The patient may exhibit visual reflexes that are mediated by lower brain centers. For example, if an object is brought rapidly toward the individual, he might blink or even move his head in an evasive way. But the subject will not be able to say why he blinked or moved his head or even that he saw something.

A lesion in one association area might lead to the loss of color vision but leave other aspects of vision normal. The patient would see objects perfectly well, moving stimuli and so forth, but everything would be colorless. A lesion still farther away from area V1 might lead to the patient's inability to recognize someone. All of this is described in detail in Chapter 8.

7

Vision

WINDOW TO THE BRAIN

··

Jonathan I. was a successful 65-year-old artist who painted abstract and colorful paintings. One day, while driving in the city, his car was hit by a small truck. He seemed unhurt but developed a severe and persistent headache, and so he went home. He slept exceptionally deeply that night, and the next morning he soon discovered he could not read, and furthermore, he noticed he was unable to distinguish colors.

A local hospital diagnosed him as having a severe concussion. His inability to read lasted for five days and then disappeared. However, his loss of color vision turned out to be permanent. For Jonathan this was devastating. His studio, filled with brilliantly colored paintings, was now only gray to him. His paintings appeared strange, and they had lost their meaning. He became extraordinarily depressed.

Neurologists surmised that Jonathan had suffered a small stroke in that part of the cortex concerned with color vision processing. The rest of his vision was unimpaired. He could see objects as sharply as before, make accurate judgments of gray scales, and see moving objects perfectly well. He could read and draw accurately, but everything was without color. He could not even visualize them in his mind!

Jonathan slowly began to paint again, and to do sculpture—but all in black and white. He also became more of a night person, feeling comfortable in the night world of muted colors. He arose later and later, to work and to enjoy the night, when he felt equal or even to some extent superior to normal people. Gradually his sense of loss left him, and he even came to feel "privileged"—that he saw a world heightened in form, without the distraction of color.

—Adapted from Oliver Sacks, *An Anthropologist on Mars*
(New York, NY: Alfred A. Knopf, 1995)

My area of research is vision and has focused on the retina of the eye, a part of the central nervous system displaced into the eye during development. I began research as an undergraduate at Harvard, working in the laboratory of George Wald, who in the 1930s discovered the role of vitamin A in vision as a key component of the light-sensitive visual pigment molecules. For this groundbreaking research, Wald received a Nobel Prize in 1967.

Initially in Wald's laboratory I studied photoreceptors and what happens to them when deprived of vitamin A. After I left Wald's lab and set up my own laboratory, my research interests moved to how the retina works, as a model piece of the brain. How do the second- and third-order cells in the retina process signals from the photoreceptors? Substantial visual processing occurs already in the retina and is reflected in the visual message coming out of the eye by way of the optic nerve.

The retina communicates much more than simply that a light is on. Indeed, since H. K. Hartline's experiments on the frog retina in the late 1930s, it had been known that some optic nerve axons respond when light illuminating the retina goes on (they give an ON response), others after light goes off (OFF response), and others briefly at the onset and cessation of light (ON-OFF response) (Figure 7.1). Hartline also showed that individual optic nerve axons respond to stimulation of only a small region of the retina, termed the *receptive field*. Stimuli presented outside the receptive field, which typically is about 1 millimeter in diameter, usually evokes no response.

We began our studies of retinal processing by examining how the retina is wired. What do the cells look like, and how do they connect with one another? Once we understood a bit about how information flows through the retina, we turned to recording the responses. This afforded insights into how cells code information and the roles of the different cells. Combined with anatomical data, the physiological results permitted us to propose schemes for how the retina is organized. We could intelligently speculate

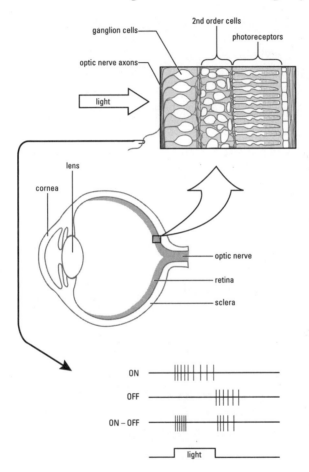

FIGURE 7.1 *The retina, consisting of photoreceptors, second-order cells, and ganglion cells, lines the back (sclera) of the eye. The axons of the third-order ganglion cells run along the surface of the retina, forming the optic nerve where they exit the eye. Light enters the eye via the transparent cornea and is focused on the retina by the cornea and lens. Recordings from single optic nerve axons (bottom) show three basic types of responses when the retina is diffusely illuminated. Some axons respond when the light is ON, others when the light is turned OFF, and some at both ON set and OFF set of the light.*

on how ON and OFF responses are generated, how the basic receptive field organization of the optic nerve's axon is established, and how complex visual information, like movement, is detected by retinal cells.

Next we began to ask how neurons talk to one another, what substances are involved, and how communication between cells can be modulated—not to mention the more recent question of how the retina develops, or what causes cells in the developing forebrain to form an eye. All our work has focused on the retina and the earliest stages of vision, but the act of seeing enlists many brain structures. From the eye, visual information passes through the thalamus in the middle of the brain into the cortex, where as many as thirty to forty different areas may participate in visual processing. Somehow it all comes together so we live in a coherent and visually rich world. But how? We have begun to glimpse underlying mechanisms, which is what this chapter is about—current ideas on retinal processing and the beginning of cortical processing. It all has relevance for understanding brain mechanisms.

Early Processing of Visual Information: The Retina

The retina lines the back of the eye and is a thin layer of tissue that consists of four major classes of neurons in addition to the photoreceptors (Figure 7.2). Within the retina two levels of processing happen: one between the photoreceptors and second-order bipolar (B) and horizontal (H) cells, and the other between second- and third-order amacrine (A) and ganglion (G) cells. The third-order *ganglion cells* are the retina's output cells; all visual information passes from the eye to the rest of the brain by way of the ganglion cell axons that make up the optic nerve.

The nature of retinal processing can be deduced by recording from the optic nerve axons—which means listening to the message being sent from the eye to the rest of the brain (see Figure 7.1). Two basic kinds of messages are being transmitted: one reflects primarily outer retinal processing; the other, inner retinal processing. Let's first consider outer retinal processing and the response properties of the ganglion cells that send on outer retinal information. Half are *ON-center cells* that respond vigorously when their receptive field center is illuminated, and the others

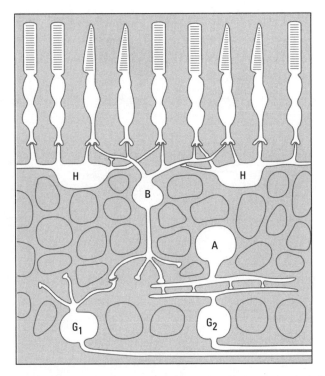

FIGURE 7.2 *The cellular organization of the retina. Photoreceptors (top) activate horizontal (H) and bipolar (B) cells in the outer part of the retina. Bipolar cells carry the visual signal from the outer to the inner retina, whereas the horizontal cells mediate lateral inhibition in the outer retina. The bipolar cells activate ganglion (G) and amacrine (A) cells in the inner retina. Some ganglion cells receive much of their input directly from bipolar cells (G_1 cell, left), whereas others receive most of their input from amacrine cells (G_2 cell, right). The ganglion cell axons run along the inner surface of the retina (bottom).*

are *OFF-center cells* that respond vigorously when light is turned off in the receptive field center. The responses of such cells are shown in Figure 7.3. Thus, two pathways are established in the outer retina; one carries ON information and the other OFF information. What might this mean?

Experiments with animals have shown that when the ON-center cells are incapacitated, the animals can no longer tell if a spot of light is intensifying or brighter than the background, but they do know if it is weakening or dimmer than the background. These data suggest that ON-center cells tell the rest of the brain about brighter illumination, whereas

OFF-center cells signal diminished illumination. Thus, whether a light spot is brighter or dimmer than the background is signaled by different pathways. From these and a slew of other experiments comes a recurrent theme with regard to visual processing: *cells and pathways within the visual system are concerned with one or another specific aspect of the scene being viewed.*

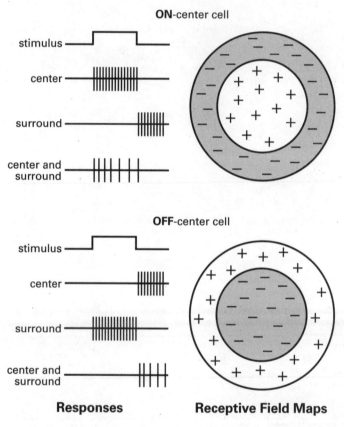

Responses   **Receptive Field Maps**

FIGURE 7.3 *Responses (left) and receptive field maps (right) of an ON-center ganglion cell (top) and OFF-center ganglion cell (bottom). The response of the cell depends on which part of the receptive field is illuminated. For an ON-center cell, illumination of the receptive field center results in an ON response (+ symbol on the receptive field map); illumination of the surrounding region results in an OFF response (– symbol on the receptive field map). When both center and surround are illuminated, a weak center-like response is evoked in the cell. The OFF-center cell behaves similarly but in a mirror-image fashion: central illumination gives rise to an OFF response; surround illumination to an ON response; and joint center and surround illumination to a weak OFF response.*

But examine Figure 7.3. More is going on with these ganglion cells than simple responses to changes in illumination. Indeed, a cell's response depends on where a spot of light falls within the cell's receptive field. Centrally the cells respond with either an ON or an OFF response, but when the light is positioned in surrounding regions, the opposite response is elicited: ON-center cells have an OFF surround, and OFF-center cells have an ON surround. Furthermore, the center and surround regions inhibit one another, so when both regions are illuminated simultaneously, the response is weak.

The antagonism between the receptive field's center and surround regions can be accounted for by a reciprocal lateral inhibition in the outer retina, similar to the reciprocal lateral inhibition between optic nerve axons in the horseshoe crab eye. In the vertebrate retina, an inhibitory neuron, the *horizontal cell*, mediates this inhibition in the retina's outer region. Horizontal cells extend processes widely in the outer retina. They receive input from photoreceptors and make inhibitory synapses on nearby photoreceptors and second-order cells (*bipolar cells*) that carry the visual message from the photoreceptors to ganglion cells. So ganglion cells with center-surround antagonistic receptive fields tell the rest of the brain not only whether light is increasing or decreasing in intensity but also something about the distribution of light on the retina. Furthermore, the simple lateral inhibition in the outer retina can explain Mach bands, the neurally mediated enhancement of edges and borders that plays a crucial role in form detection discussed in Chapter 5. Finally, we see the beginning of color processing in the outer region of many retinas; that is, the centers of the receptive fields of certain ganglion cells in animals with good color vision are maximally responsive to light of one color— they receive their input from just one cone type—whereas the receptive field surrounds are most responsive to light of another color. Thus, much processing goes on in the outer retina, and this is just the beginning!

In the inner retina, between the second- and third-order cells are neurons that detect movement. When a spot of light is shone on the retina and left there, many of these neurons, called *amacrine cells*, initially respond vigorously, but then their responses rapidly fade away. When the light is turned off, these cells also respond, but again the responses quickly fade away. By contrast, moving the spot around on the retina elicits vigorous activity in the amacrine cells that lasts as long as the spot remains in the receptive field (Figure 7.4a). This movement sensitivity is passed on to a second group of ganglion cells, the *ON-OFF cells*, which are highly movement sensitive and in some animals are even direction sensitive (Figure 7.4b): a spot of light moving across the retina in one direction elicits activity, but the same light moving in the opposite direction excites no response.

So, from the eye to the rest of the brain go two basic messages—one reflecting mainly outer retinal processing, the other inner retinal

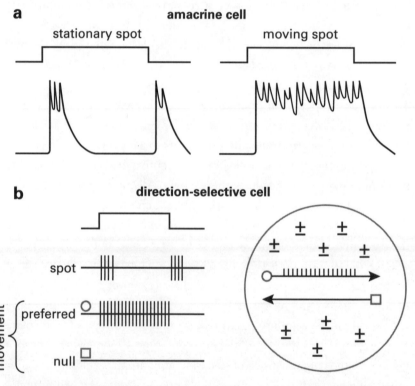

FIGURE 7.4 *Amacrine cell responses (a) and responses and a receptive field map of a direction-selective ganglion cell (b). To a stationary spot of light, amacrine cells respond with a transient response at the onset and offset of the light. To a moving spot, the cell responds continuously. A directionally selective ganglion cell responds with an ON-OFF response to a stationary spot of light projected anywhere in the receptive field, indicated on the map by a ± symbol. When a spot of light is moved through the field in the preferred direction, vigorous activity is elicited in the cell. When the spot moves in the null direction, no activity is elicited.*

processing—by way of two sets of cells. One set (G$_1$ in Figure 7.2) provides luminance, spatial, and (sometimes) color information; the other (G$_2$ in Figure 7.2) provides temporal information about the light impinging on the retina. The latter cells respond vigorously to movement and the visual image's dynamic properties, which leads to the second recurrent theme in visual and brain processing: *information is processed along parallel pathways, and abundant parallel processing occurs within the brain.* So, different aspects of the visual image, such as form, color, and movement, are analyzed simultaneously along different pathways. This is a key feature of how the brain works and why our brains are far superior to any computer so far devised. We can process many things simultaneously that require neural computations: talking, walking, seeing, hearing, touching, and so forth. Computers process information serially for the most part; they do one computation at a time. Parallel-processing computers are beginning to emerge, but such computers are difficult to program. Thus, even though the processing components that make up our brains, the neurons, operate slowly compared with the electronic units in computers, our brain far outstrips any computer so far devised because of our ability to parallel process in a massive way.

Next Stages of Visual Processing: The Primary Visual Cortex

From the retina, the visual messages go to the cerebral cortex by a way station in the thalamus, the *lateral geniculate nucleus (LGN)*. (Figure 7.5) The receptive fields of LGN neurons are very similar to those of the ganglion cells, indicating that relatively little new processing of visual information occurs there. However, the LGN receives substantial input back from the cortex and also from the reticular formation in the medulla. Thus, visual information can be gated—strengthened or weakened—in the LGN before it goes on to the cortex. Axons from the LGN then project to the occipital lobes of the cortex, where the visual signal is first processed at the back of the occipital cortex in the primary visual area, area V$_1$ (V is for "visual").

When neurons in *area V$_1$* are recorded, and stimuli presented to the

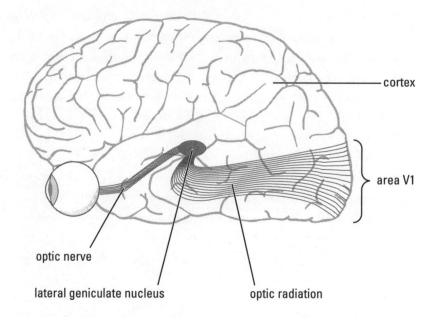

FIGURE 7.5 *The central visual pathways. The optic nerve extends from the eye to the lateral geniculate nucleus in the thalamus. From the lateral geniculate nucleus, the visual message goes to the primary visual area (V1) of the cortex via the optic radiation, which consists of the axons of lateral geniculate neurons.*

retina, a considerable elaboration of receptive field organization happens. Rather than responding vigorously to circular spots of light, either stationary or moving, presented to the retina, the cortical cells require more complex stimuli to respond maximally. In area VI, several types of neurons are recorded. *Simple cells* are found closest to the input areas in the cortex (Figure 7.6). They respond best to bars or edges of light that are oriented at a specific angle. Some cells respond best when the bar or edge is straight up or down, but others when the bar or edge is on a slant, at 45 degrees, for example. Still other cells respond best to horizontal bars or edges. Altering the optimal orientation by about 10 degrees causes the cell to fire less well.

Complex cells are found farther away from the input areas of the cortex. They also respond best to highly oriented bars of light, but the bar must be moving, and moving in a specific direction—at right angles to the orientation of the bar (Figure 7.6). These cells respond only weakly

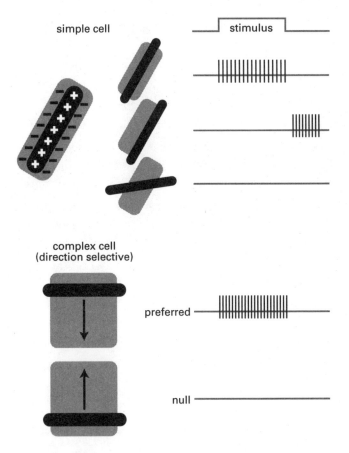

FIGURE 7.6 *Receptive field maps (left) and responses (right) for simple (top) and complex (bottom) cortical neurons. The simple cell responds best to an oriented bar of light that fits its central excitatory zone (+ symbols). Moving the bar into the surround region (– symbols) elicits an OFF response from the cell. Stimulating the field with an inappropriately oriented bar of light results in little or no response. The complex cell responds to an oriented bar of light or dark moving at right angles to the bar's orientation. The cell illustrated is direction sensitive. Movement in the preferred direction elicits vigorous activity; in the null direction, no activity.*

to a stationary bar of light, even when the orientation is correct. More specialized complex cells are also recorded, and they are located even more distally from the input areas. Some of these cells have *direction-selective* properties. Bars moving in one direction vigorously excite the cells, but bars moving in the opposite direction inhibit the cells. Stationary bars of light projected on the retina cause little or no response, as do bars not moving at right angles to the preferred direction of movement. Other specialized complex cells require moving bars that are restricted in their length. Bars that are longer than optimal begin to inhibit the cells' activity.

The main conclusion drawn from these observations is that the farther along the visual system one records a neuron, the more specific must be the stimulus presented to the retina to drive the cell maximally. Whereas spots of light are sufficient to drive retinal ganglion cells vigorously, simple cortical cells require oriented bars or edges of light. Complex cells require not only oriented light bars but bars that are moving, and specialized complex cells require an oriented bar moving in one direction and/ or restricted in length. You can think of this as an abstraction process: specific cells respond only when a specific set of criteria with regard to a stimulus is met. Thus, when a complicated figure is projected onto the retina, the responding cortical cells are few compared to the total number of neurons receiving input from that part of the retina—only cells respond whose receptive field requirements are matched by a part of the figure. Put in another way, components of a figure are encoded by specific

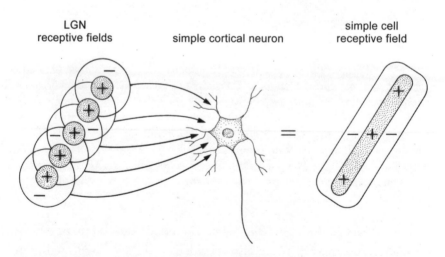

FIGURE 7.7 How a simple cortical cell's receptive field could be fashioned by excitatory inputs from the LGN. The receptive fields of the input neurons (ON-center cells) overlap and lie along a straight line on the retina (left). The receptive field of a cortical neuron receiving such input will consist of an elongated central excitatory zone surrounded by an inhibitory zone (right).

neurons. A figure is thus analyzed by the visual system, with individual neurons responding to one or another part of the figure.

Formation of Cortical Receptive Fields

Although the synaptic circuitry of the cortex is not yet worked out in detail, one can speculate as to how cortical neuron receptive fields might be structured from the LGN input neurons or from synaptic interactions among the cortical neurons themselves. For example if ON-center LGN neurons that receive input from ganglion cells that are aligned along a particular orientation on the retina and all feed into one cortical cell, as shown in Figure 7.7, that cortical cell would have the properties of a simple cell. The overall receptive field size of simple cells is much larger than the receptive field size of a single LGN neuron, which provides support for this idea.

Directionally selective complex cells could be generated from edge-selective simple cells that provide input to a complex cell. If the edge-selective

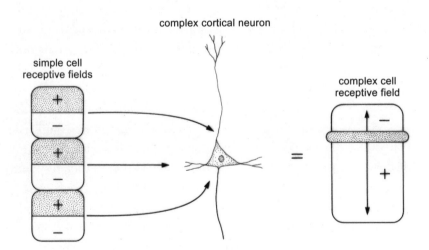

FIGURE 7.8 *How a directionally selective complete receptive field could be formed from edge-selective simple cells. If the receptive fields of the input neurons are aligned on the retina as shown on the left, downward movement of a bar will elicit vigorous activity in the complex cell, whereas upward movement will result in no response.*

simple cells are aligned as shown in Figure 7.8, a bar moving in the downward direction would encounter first the excitatory zones of the simple cells' receptive fields, and the complex cell would be strongly stimulated. A bar moving in the opposite, upward direction would activate the inhibitory zones first, thereby inhibiting the excitatory zone, and no response would occur in the directionally selective complex cell.

The schemes shown in Figures 7.7 and 7.8 are undoubtedly overly simplified, but they do provide a way to think about possible cortical synaptic organization and how the hierarchy of cortical cell types may be formed.

Seeing Depth: Binocular Interactions

Many animals, including cats, monkeys, and ourselves, have eyes that face forward, which means that the visual fields of the eyes overlap. The advantage of having two eyes whose visual fields overlap is the capability of depth perception. This is easily demonstrated by trying to oppose index figures if one eye is closed and your head is still; more often than not, you will fail this task. To accomplish this task easily and reliably, input from both eyes to the same cortical neurons is required, and most neurons in V1 receive input from both eyes—they are binocular. Indeed, it is in V1 that binocular cells are first encountered. Cells in the LGN are monocular, receiving input from one or the other eye.

It has been found, however, that cortical cells are most often driven more strongly by one eye than by the other, a phenomenon termed *ocular dominance*. This is shown in Figure 7.9, in experiments carried out in the cat. It shows the relative number of cells encountered when recording cells throughout the cat cortex as a function of whether the cell responds to only one eye (groups 1 and 7), equally from both eyes (group 4) or to different levels of input from one eye or the other (groups 2 and 3, and 5 and 6). The cells receiving input from just one eye (monocular cells) are found in the input layer of the cortex, and binocular cells, more distally; the more binocular, the farther away from the input layer. Thus, there appears to be a hierarchy of degree of ocular dominance just as there is in terms of receptive field complexity, that is, simple cells before complex cells, and so forth.

When a cell receives input from both eyes, both eyes must be stimulated in the same way, and in the corresponding retinal location. If an oriented bar of light moving in one direction is required to best activate a cell receiving input from one eye, the exact same stimulus is required to drive the corresponding cell in the other eye. If both eyes are stimulated together with the same stimulus, the response of the cortical cell is greater than if just one eye is stimulated. However, the response of the cell is usually greater from one eye than the other, showing ocular dominance (unless the cortical cell is driven equally by both eyes).

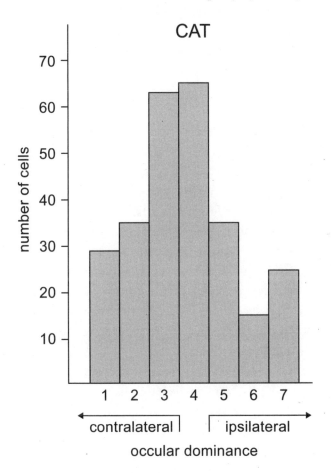

FIGURE 7.9 *Ocular dominance histograms for cortical cells in the cat. Cells in groups 1 and 7 are monocular (driven by one eye), whereas cells in group 4 are equally binocular (driven as well by both eyes). Cells in groups 2, 3, 5, and 6 are binocular but are driven more by one eye or the other.*

How does binocularity of the cortical cells relate to depth perception? Although all binocular cortical cells have their receptive fields in comparable retinal areas in the two eyes, most respond well even if the stimuli presented to the two eyes are not perfectly aligned. Certain complex cells, however, do require precisely aligned stimuli to be presented to the two eyes for the cells to respond. Termed *disparity-tuned cells*, these cells are believed to be critical for depth perception.

Because our two eyes are separated laterally, images near or far fall on corresponding but slightly different parts of the retina. For most cells in VI, this makes little difference. For disparity-tuned cells, this difference is critical. Some respond only when images are near; others, only when images are farther away. In other words, the disparity-tuned cells require exactly positioned images on the retina for them to respond optimally, and it is thought, therefore, that these cells signal the position of an object in depth.

Cortical Organization: The Hypercolumn

There is much going on in each part of area VI. There are cells with different receptive field properties, and this appears hierarchical in nature. Simple cells come before complex cells, which come before more specialized complex cells. But all of the VI neuronal receptive fields described so far (except for a few in layer 4) have a requirement for orientation. Although some VI cells receive input from only one eye, most receive input from both eyes but to varying degrees. And there are VI cells interested primarily in color, which are discussed further below.

David Hubel and Torsten Wiesel, who carried out many of the experiments on VI cortical cells (and for which they were awarded a Nobel Prize in 1981), proposed that there is a basic cortical organization in area VI that takes into account all of this complexity. That is, the cortex is organized into modules, which they called *hypercolumns*, that have the dimensions of 1 × 1 × 2 millimeters, the latter being the thickness of the cortex. This basic cortical structure contains all the neuronal machinery necessary to analyze a bit of visual space. These are not completely separate modular units, but they overlap.

The organization of a hypercolumn is shown in Figure 7.10. First, input from an eye comes into the middle of the cortex (called layer 4), on one or the other side of a hypercolumn, forming irregular stripes that run across layer 4. The columns are about 0.5 millimeter wide, so that to encompass information from both eyes requires 1 millimeter of

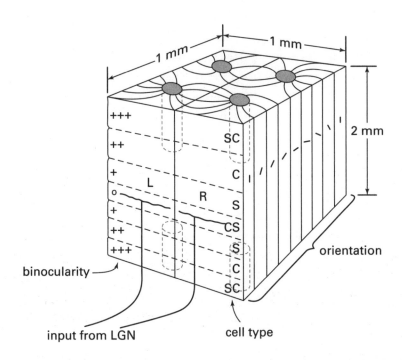

FIGURE 7.10 *A hypercolumn; a 1 × 1 × 2 millimeter block of cortex containing all the cells required to analyze a bit of visual space. In a hypercolumn, input from both eyes (L or R) is represented. Furthermore, all types of simple and complex cells are present, and the cells have all possible orientation preferences and varying degrees of binocularity. In addition, color-sensitive cells are found in the pegs inserted into the hypercolumn (represented by cylinders). Degree of binocularity is represented by + symbols; where lateral geniculate input enters the cortex, the cells are monocular (represented by an open circle); away from this layer the cells are more and more binocular. Cell types also vary through the thickness of the cortex. Around the input layer some cells are center-surround (CS); away from this layer they are first simple (S), then complex (C), and finally specialized complex (SC). Orientation-selective cells are found in narrow columns that run at right angles to the ocular dominance columns (L or R). These columns converge in the middle of the ocular dominance columns to form the round pegs where the color-preference cells are found.*

cortex—one dimension of the hypercolumn. Above and below where the LGN input comes in, the cells are binocular, the extent of which depends on how far the cell is from layer 4. Cells in layer 4 usually show no binocularity; they are monocular, as indicated by a circle on the left side of Figure 7.10. The degree of binocularity is indicated by + symbols; three + symbols indicates the cell is receiving roughly equal input from both eyes.

A similar organization is seen for cells with different receptive field properties (right side of the hypercolumn in Figure 7.10). Many cells in layer 4 have a center-surround receptive field organization (CS), whereas simple cells (S) are found above and below these cells. Complex cells (C) are found farther away from the LGN input, and specialized complex cells (SC) farther still.

Running roughly at right angles to the ocular dominance columns are orientation columns. There are eighteen to twenty or so different orientations that cells can prefer, and these columns are much narrower, about 0.05 millimeter in width. To take into account all the orientation possibilities requires again about 1 millimeter of the cortex, the second dimension of the hypercolumn. The orientation columns are clearest at the edges of the ocular dominance columns. This is because in the centers of the ocular dominance columns are groups of cells that have color preference, and the orientation columns converge into the color areas as shown in Figure 7.10. The grouping of color cells does not go all the way through the thickness of the cortex; rather, they are found above and below layer 4 and often are referred to as pegs or blobs. The cells in the color pegs have a center-surround receptive field organization and may be either single or double opponent: that is, red sensitive in the center, green sensitive in the surround (single opponent); or red (ON-response), green (OFF-response) in the center, and opposite red-green responses in the surround (double opponent) (Figure 7.11). They are the only cortical cells outside of layer 4 not to have an orientation preference.

The cortex, then, consists of a series of repeating modules, each looking after a bit of the visual field. Whereas the major organizing principle of the cortex is vertical, horizontal connections between the modules also link cells within a specific layer. This means, for example, that infor-

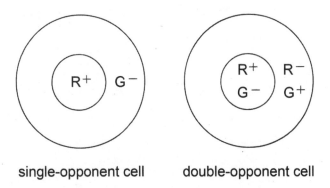

single-opponent cell   double-opponent cell

FIGURE 7.11 *Receptive fields and single- and double-opponent color cells. R⁺ means that red light excites that region of the receptive field; G⁻ means that green light inhibits that region of the field.*

mation is integrated over many millimeters of cortex, and thus a cell in one module can be influenced by stimuli falling on other modules. A revealing example is shown in Figure 7.12, where the orientation of lines appears altered by the orientation of surrounding lines. Known as the tilt illusion, the lines in the center appear tilted to the left, although they are perfectly vertical.

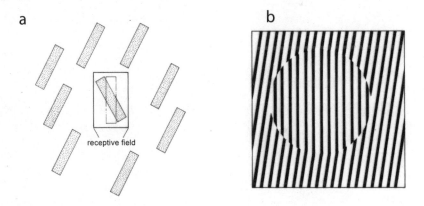

FIGURE 7.12 *The effects of distal stimuli on the response of a cortical neuron. (a) The orientation tuning of a cell can be altered by up to 10 degrees by the presence of orientated stimuli outside the receptive field. The orientation selectivity of the neuron (clear bar) is tilted to the left (shaded bar) when bars of light tilted to the right are presented outside the receptive field. (b) The tile illusion. The lines in the circled area appear tilted to the left, although they are vertical.*

PART THREE

Cognitive Science

HIGHER BRAIN FUNCTION
AND MIND

The ultimate goal of neuroscience is to understand higher brain function and behavioral phenomena we often refer to as mind or cognition. Perception, language, memory, learning, emotion, motivation, and consciousness are all aspects of mind and are considered in this last section. Our understanding of these phenomena is still rudimentary, but many of them can be localized to specific brain areas, and in some cases we have even gained some insights into their underlying cellular and molecular mechanisms. Whereas in the past separation of mind and brain was vigorously argued by distinguished philosophers and scientists, virtually everyone today believes that mind originates in brain function and is an emergent property of the brain.

8

From Brain to Mind

VISUAL PERCEPTION

Dee was a freelance commercial translator who lived in Italy with her partner, Carlo. She was an active and lively person, but one fateful day she had a tragic accident that changed her life permanently.

Dee was taking a shower in the house she and Carlo had newly bought. The shower was heated by a propane gas heater, but the heater was improperly vented and carbon monoxide slowly accumulated in the bathroom. Unable to detect the odorless fumes, she collapsed into a coma as the carbon monoxide displaced the oxygen in her blood. Carlo arrived home just in time to save her, rushed her to the local hospital, and she survived.

When she regained consciousness she could speak and understand things but could see nothing. Over the next few days, Dee began to regain some conscious sight—first color, then surface features, like the tiny seeds on a strawberry, indicating her visual acuity was excellent. However, she could not recognize objects, faces, or words, and her vision did not improve.

Dee had, however, retained surprising visual abilities that were revealed one day when she was asked if she could recognize a pencil held in front of her. She could not, but then reached out and skillfully grasped it. Upon further testing, she could accurately grasp the pencil regardless of its orientation or location in front of her. She obviously had visual abilities but was not conscious of them.

Further testing demonstrated other examples of her amazing visual abilities. Her testers wrote:

The contrast between what she could perceive and what she could actually do with her sense of vision could not have struck us more forcibly than it did one day when a group of us went out on a picnic while visiting. We had spent the morning at her home carrying out a series of visual tests, recording one failure after another.

To lighten the gloom, Carlo suggested that we all go for a picnic in the Italian Alps, to a popular spot not far from their home. We drove high up into the mountains and then set off on foot to reach our picnic site—an alpine meadow. To reach the meadow, we had to walk along a half-mile trail through a dense pine forest. The footpath was steep and uneven, yet Dee had no trouble at all. She walked confidently and unhesitatingly, without stumbling, tripping over a root, or colliding with the branches of trees that hung over the path. We had to point out to her the correct route to take, but other than that, her behavior was indistinguishable from that of any of the other hikers on the mountain that day.

We arrived at the meadow and began to unpack the picnic hamper. Here Dee displayed once more how apparently normal her visual behavior was. She reached out to take things that were passed to her with the same confidence and skill as someone with completely normal sight. No one would have guessed she could not see the difference between a knife and a fork, or recognize the faces of her companions.

—Adapted from Melvin Goodale and David Milner, *Sight Unseen* (Oxford, UK: Oxford University Press, 2004)

In Chapters 6 and 7 I discussed the primary visual cortex, area VI. What about other primary sensory areas? How are they organized? Present evidence suggests that other primary sensory areas, serving the somatosensory and auditory systems, are organized similarly to the primary visual cortex. For example, neuroscientists have observed a columnar organization in the primary somatosensory and auditory cortex. In fact, the first columns in the cortex were discovered in the somatosensory cortex by Vernon Mountcastle at Johns Hopkins University in the 1950s.

The columns in the primary somatosensory cortex encompass cells that respond to different types of stimuli. Cells in one column are responsive to light touch; cells in other columns respond to deep pressure, to the movement of hairs, or to joint position. The neurons in a column all receive their input from the same areas of the skin or limb, and their receptive field areas overlap. Stimulation within the receptive field either excites or inhibits the firing of the neurons. The receptive fields of many neurons in the somatosensory cortex are also organized in ways reminiscent of the visual system.(Figure 8.1) The receptive fields of tactile neurons may have an antagonistic center surround organization. The center of the cell's receptive field is inhibited by simultaneous stimulation in the periphery of the receptive field. In the region of the somatosensory cortex receiving input from the hands, a few neurons respond much better to moving tactile stimuli than to static touch, and some of these neurons show a preference for direction of movement—they are direction-sensitive.

In the primary auditory cortex, neurons in a column tend to respond to similar frequencies, and there is a progression along the primary auditory cortex of the frequencies to which the neurons best respond. The neurons in a column may have different temporal properties: some respond transiently to a brief sound; others, in a sustained fashion. Other neurons appear to be modulated by intensity—they respond best to a particular intensity of sound.

Some neurons in a column typically respond only to sharply tuned stimuli, whereas others respond well to complex sounds like clicks, and they show minimal tuning. The frequency columns, like the ocular dominance columns in the visual cortex, are organized as stripes or bands that run across the primary auditory cortex.

Another interesting feature of the auditory cortex's organization concerns the input from the two ears. Some neurons respond best to stimulation of both ears, whereas others respond best to stimulation of one ear and are actually inhibited if both ears are stimulated. The cells are also organized in columns, which are thought to be at right angles to the frequency columns or bands. Thus, even the architecture of the auditory cortex resembles that of the visual cortex. Spreading beyond the primary (cortical) areas are secondary (or association) areas of the cortex. As noted earlier, these areas are concerned with more complex processing, with integration of sensory modalities, and ultimately with recognition, understanding, and memory. How are these

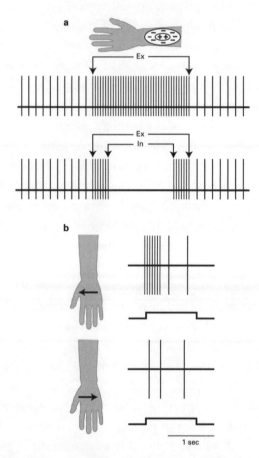

FIGURE 8.1 *Receptive fields of tactile neurons recorded in the somatosensory cortex. (a) Many of the neurons have an antagonistic center-surround receptive field organization. Stimulation of the center of the field (Ex) excites the cell; surround stimulation (In) inhibits firing of the cell. (b) Some neurons are movement and even direction sensitive. Movement of a tactile stimulus in one direction vigorously activates the cell; movement in the other direction only weakly excites it.*

areas organized? What do we presently know about them? Let us return to the visual system, where we have more information, and focus on visual processing beyond areas V1. I then discuss present ideas on visual perception.

Area V2 and Higher Visual Areas

The main target for the axons leaving area V1 is the adjacent *area V2*. Clues to the organization of V2 came initially from anatomical studies of cortex stained for an enzyme (cytochrome oxidase), but the significance of this staining is not understood. Tangential sections of area V2 stained for this enzyme showed a pattern of alternating dark- and light-staining stripes (Figure 8.2). Further, the dark stripes are of two types, thick and

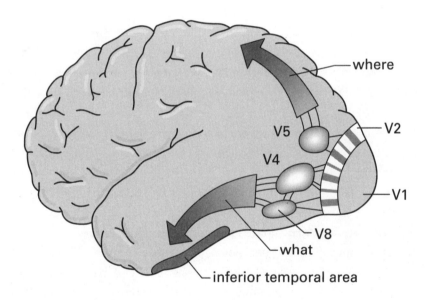

FIGURE 8.2 *Visual areas beyond V1. Enzyme staining shows banding of area V2 into thick bands, thin bands, and interband regions, where different aspects of the visual image are processed. Beyond V2, separate areas are concerned with processing these aspects of the visual image: V4, form and color; V8, color; and V5, movement. Visual information progresses to yet higher visual areas via two pathways: a dorsal pathway into the parietal lobe known as the where pathway and a ventral pathway into the temporal lobe known as the what pathway. A face recognition area is localized on the underside and inner surfaces of the temporal lobes (inferior temporal area).*

thin. Thus, across V2, a regular and repeating pattern occurs of thick and thin dark stripes separated by light stripes.

The obvious question raised by these observations is whether cells in these subdivisions have different properties, a question examined by Margaret Livingstone and David Hubel, who found surprising differences. Cells in the light stripes were often color coded. Most were double-opponent cells and had properties not seen in area V1, suggesting further processing of color information. Cells in the thin dark stripes were orientation selective but not direction selective. These cells appeared concerned mainly with image shape. In the thick dark stripes, most of the cells were directionally selective and highly binocular. That is, they responded poorly to stimulation of just one eye but vigorously when both eyes were stimulated simultaneously. These cells therefore appeared to be concerned primarily with movement and depth.

The conclusion drawn from these findings is that further segregation of visual image processing exists in area V2. Although the stripe subdivisions are interdigitated in area V2, explorations of visual areas beyond V2 show that the segregation becomes complete. In other words, separate regions of the cortex become specialized for one or another kind of processing (see Figure 8.2). *Area V4*, for example, appears to be concerned mainly with form processing and perhaps some color discriminations. It receives input from both the light and thin dark stripe areas of V2. An area anterior and ventral to V4 (sometimes called *area V8*) seems to be specialized for color processing; it is strongly activated by colored stimuli. *Area V5* or MT (middle temporal) is involved in the analysis of movement and perhaps depth. For example, the responses of neurons in V5 correlate closely with the ability of monkeys to discriminate direction of movement. V5 receives much input from the thick stripes of V2.

From examining areas V4, V5, and V8, we know that each has a representation of the retina—the retina is topographically mapped onto these areas. Thus, distinct areas analyze one or another aspect of the visual image. There are consequences to this way of analyzing a visual image. For example, because form and color are processed separately for the most part, form is difficult to see if an image is made from two colors that have the same brightness. To see form well, we need light-dark contrast; color contrast by itself will not do. Another consequence of this arrangement—and one that also provides additional evidence for it—is that certain brain lesions can affect a person's ability to see color but no other aspect of the visual image, as was the case for Jonathan I. described at the beginning of Chapter 7. Such individuals may see shapes perfectly well, have high visual acuity, respond to moving stimuli, and have good depth perception, yet they see no colors. These symptoms suggest that the lesion is confined to area V8, as may occur after a small stroke.

People with other brain lesions have been known to lose the ability to see movement, even while other aspects of vision—such as perception of colors, static objects, and depth—are preserved. Loss of movement vision is severely debilitating; such patients cannot cross a street unaided because they cannot tell whether cars are moving toward them. A woman who had a stroke that involved area V5 described what it is like: "When I'm looking at the car, it first seems far away. But then, when I want to cross the road, suddenly the car is very near." Pouring a cup of tea is also extremely difficult for her, for the liquid flowing from the pot looks solid to her, and she cannot see the tea rising in the cup.

In addition to those areas mentioned above, perhaps as many as twenty-five to thirty additional areas are involved in analyzing one or more aspects of the visual image. These areas participate in two major pathways of visual information flow. One pathway progresses from the occipital lobe dorsally into the parietal lobe. Known as the *where pathway*, it provides information on where an object is in space and is involved in the visual control of reaching and grasping. The other pathway flows from the occipital lobe ventrally into the temporal lobe. It is known as the *what pathway*; it identifies objects. Individuals with damage to the *what* pathway may say they cannot recognize an object but are able to reach out and grab hold of it accurately regardless of its orientation. Dee, the victim of carbon monoxide poisoning described at the beginning of this chapter, is such an example. She could not recognize objects but could grasp objects accurately and even navigate in space. Presumably her *where* pathway was unaffected. Those with lesions affecting the *where* pathway would recognize objects but have difficulty finding and grasping them, and would presumably fail at navigating.

Face Recognition

A fascinating ability that humans and some animals have is to recognize faces rapidly and reliably. In the early 1970s Charles Gross and his colleagues at Princeton University found some cells in the inferior temporal cortex of monkeys that respond selectively to faces. It had been known from clinical studies that people with lesions in this area often could no longer identify people by their appearance, something that normal individuals do easily. Deficits in recognizing things are known as *agnosias*; failure to recognize faces is termed *prosopagnosia*. Some people without known brain lesions also cannot recognize faces. Oliver Sacks, the well-known author who wrote on many neurological deficits, was one of those individuals. His brother also was prosopagnosic, suggesting it can be a genetic defect. Other than being unable to recognize people visually, these people are usually remarkably free of other visual deficits—they can read, write, name objects, and so forth, but they cannot visually recognize even a very familiar person, including a spouse. They can't even recognize their own face in a picture or mirror, yet they know it is a face they are looking at. They can recognize a familiar person by voice and immediately name them, and can describe a face in detail and the emotion shown by a facial expression, but they cannot associate a face with a person.

Exactly what is involved in face recognition is not well understood. For example, it is difficult to recognize a face that is upside down. Furthermore, faces that have been distorted do not appear nearly as peculiar when viewed upside down as when viewed right side up. When viewed upside down, the two faces in Figure 8.3 appear more or less equivalent. When viewed the other way around, the two faces are dramatically different.

Recordings from neurons in the monkey's inferior temporal area, comparable to the face-recognition area in humans, have yielded intriguing results. The neurons in this region of the cortex have large receptive fields, and most of them include the *fovea*, which means that the cells are concerned with the central visual field. Most interesting is the finding that about 10–20 percent of the cells respond best to complex images, including images of faces and hands. If the image is simplified—made

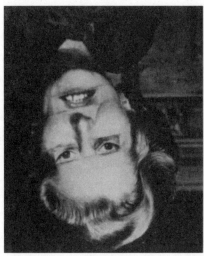

FIGURE 8.3 *Upside-down pictures of Mrs. Thatcher. The two pictures appear approximately equivalent when viewed upside down, but when turned right side up they appear very different. In one of the photographs, they eyes and mouth have been inverted.*

less detailed by blanking out the eyes in a picture of a face, for example, or by filling in the space between the fingers—it elicits a weaker response from the neurons than does the original.

Figure 8.4 shows a recording from a neuron that responded better to faces than to other complex stimuli. The neuron responded best to the picture of an intact face (a); when the eyes in the picture were blanked out, the neuron responded less vigorously (b). When the picture was cut into sixteen pieces and randomly rearranged, the neuron didn't respond at all (c). It also did not respond to a hand (d). But other neurons in the area did respond selectively to hands and other complex objects, so the neurons in this area are not exclusively face cells.

The finding that individual cells respond selectively to faces and other complex objects has aroused much interest among neuroscientists, and face recognition has recently been intensively studied. Six areas (called patches) have now been implicated in face recognition, and how individual cells recognize faces is now being explored. Some face cells respond more readily to certain facial features, such as eyes or hair, roundness of a face, distance between the eyes, and so forth.

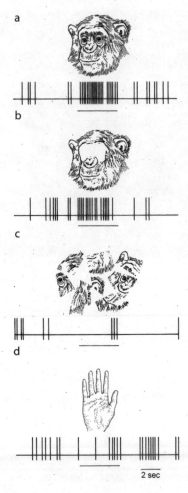

FIGURE 8.4 *Recordings from a monkey inferior temporal neuron that responded best to faces. The images were projected onto the retina for 3 seconds; the records below indicate the resulting action potentials. The neuron responded best to the intact face (a), less well to an image with the eyes blanked out (b), and not at all to a picture of a face cut into pieces and rearranged (c) or to a hand (d).*

Overview of Visual Processing

Two general themes have emerged from the study of visual processing at all levels of the visual system, from the retina to visual areas V4, V5, V8, and beyond. These themes offer insights into how we perceive images.

The first, already pointed out in Chapter 7, is that cells and pathways are concerned with the processing of one or another aspect of the visual image. Already at the level of the photoreceptor synapse, we find that ON- and OFF-information is segregated into two classes of bipolar cells; this segregation is maintained throughout the visual system. Furthermore, the outer retina is concerned primarily with spatial aspects of a visual image as well as color, while the *inner retina* emphasizes more the temporal and complex aspects of a visual signal. Two basic classes of ganglion cells providing information about these fundamental aspects of the visual scene project to higher visual centers. Again, segregation of these aspects of the visual image is maintained in the lateral geniculate nucleus and throughout the cortex. Beyond visual areas V1 and V2, more specific components of the visual image—including form, color, and movement—are dealt with separately and in separate cortical areas. There exist, then, *parallel processing* streams in the cortex that give rise, ultimately, to the relatively unified visual world we perceive.

A second common theme is that the visual system is not designed to make absolute judgments but, rather, to make comparisons. This characteristic is also seen first in the retina and is reflected in the receptive field organization of bipolar and ganglion cells. That is, the receptive fields of bipolar and ganglion cells consist of antagonistic center-surround or color-opponent regions. It is not the absolute intensity of light that comes from an object that makes it appear light or dark; it is the intensity of light coming from the object relative to the intensity of light coming from surrounding objects. (Figure 8.5) Two practical examples also serve as illustration. When turned off, a television screen appears gray (at least older TVs). When turned on, the television picture displays black as well as all shades of gray and bright white. There is no such thing as negative light; the natural gray of the TV screen appears black when the set is on because of the lighter areas adjacent to it. Reading a newspaper in very dim or very bright light is another example. In either case, the print appears black and the rest of the paper white regardless of the surrounding illumination level. But if you measure the light reflecting off the black print in bright sunlight, it can exceed the intensity of light reflecting off the white areas in dim light! Not only does perceived brightness depend on the surround, but an object's perceived color also depends on the color of surrounding objects. No one picks out

FIGURE 8.5 *The bar is uniformly gray from one side to another but appears lighter against a dark background, while a lighter background makes it appear to be darker.*

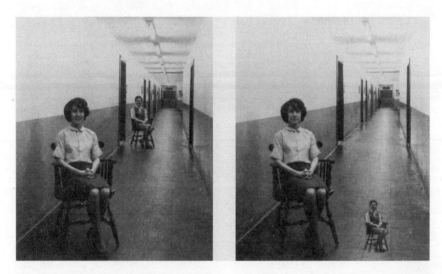

FIGURE 8.6 *Context affects the perception of size; the scene was photographed twice, once with both women present, once with only the woman in the foreground. The image of the other woman was cut out from a print of the first picture and pasted onto the second picture. She now appears much smaller than in the first picture, but she is exactly the same size in both photographs (measure them yourself!).*

upholstery material for curtains, chairs, or couch without taking a swatch home to see if it looks right in the designated space. Surround is even critical for perceived judgments of size. Figure 8.6 is a picture of two women. They appear of comparable size in the picture on the left, but they are seen as vastly different in the one on the right simply because of the position of the "smaller" woman. Yet her image is the same size in both pictures.

Visual Perception

The retinal image is two-dimensional, yet we live in a three-dimensional world and perceive it so. Our visual system uses many pieces of information to reconstruct an image. If not all the pieces are consistent, or if they are not all there, the system nevertheless attempts to provide a complete and coherent picture. The system can be confounded, however, as anyone who has seen a visual illusion is aware. Figure 8.7 displays examples of illusionary borders. Although no lines define the shapes, most people will easily make out a well-demarcated triangle and circle in the top two figures and a contoured line running through the bottom figure. Furthermore, the circle and triangle seem brighter than the surrounding white areas. Visual perception is therefore reconstructive and creative. This was appreciated in the beginning of the twentieth century by the Gestalt psychologists, who put forward the idea that the brain creates an image by taking in information concerning the components of a visual scene and by making certain assumptions about the shapes, colors, and movements within the scene—all are needed to create a coherent visual picture. If something is missing from the scene, it is filled in. Each of our retinas has a hole in it, where the optic nerve exits. But when we look out at the world we do not have a hole in our field of view. The brain fills in that defect so that we don't usually notice it. Yet this blind spot can be demonstrated by a simple exercise, as in Figure 8.8. If the image we are looking at is ambiguous, we perceive one thing or another but not a mixture of two percepts. An example of the latter phenomenon is the famous vase-face illusion shown in Figure 8.9. At any moment we may see two black faces in the picture or a white vase, and it is fairly easy to switch back and forth between the two percepts. We see only one of the two percepts at a time, however.

How images are re-created in the cortex is not well understood and is known as the binding problem. Interactions between the processing streams must occur, perhaps at several levels. Both psychological and physiological experiments have emphasized the role of attention in extracting information from the processing streams and in visual perception. If there are multiple objects in the visual scene, it is possible to focus on only one or a few objects at a time; perception of the other objects in the field of view is actively suppressed. To pick out an object from the surround in a scene requires that specific attention be paid to it, as exemplified by the effort needed to switch from one percept to another in the face-vase illusion. This mechanism may help filter out irrelevant material in a complex visual scene.

FIGURE 8.7 *Examples of illusionary borders. Although the shapes are not drawn as such, a well-defined triangle and circle appear in the top two figures. Note also that the circle and the triangle appear brighter than the surrounding white areas, but they are not. In the bottom figure, a curved line separating the two halves of the figure appears to be visible.*

FIGURE 8.8 *The physiological blind spot. Close your left eye and focus your right eye on the X. Move the book slowly toward or away from your face and the butterfly on the right will eventually disappear. It reappears when the book is twisted or moved one way or the other.*

FIGURE 8.9 *The face-vase illusion. At any one time two black faces or a white vase is perceived. Never are the two percepts seen simultaneously, although some people can switch percepts so quickly they think they can see them simultaneously.*

Modulation of Cortical Responses

The picture developed so far may suggest that cells along the visual pathways and the rest of the brain are quite fixed in their properties. This is not strictly true. We know, for example, that stimuli presented far outside the normal boundaries of a conventionally mapped receptive field of a retinal ganglion cell or lateral geniculate nucleus neuron can affect the responses of the cell. Moving a dark spot vigorously as much as 10 milli-meters away from the receptive center of such a neuron will not activate the cell, but it may increase or decrease the responses of the cell to spots of light falling on the cell's receptive field. We suspect that such long-range retinal effects are mediated by certain types of amacrine cells that extend processes long distances laterally in the inner plexiform layer of the retina. Similar and even more dramatic effects are seen in the visual cortex. By presenting specific stimuli outside cortical receptive fields, it is possible not only to alter the responsiveness of the cortical cells but also to alter aspects of receptive field organization, including receptive field size, degree of ocular dominance, and even orientation selectivity. Axons, which extend laterally in the cortex for distances of over 6 millimeters, may be responsible for mediating these long-distance horizontal effects. It is believed that these long-range lateral interactions occurring within the cortex are responsible, at least in part, for contextual effects that are at play in visual perception.

The classic receptive field is determined with light spots or bars pre-sented to the retina that by themselves excite or inhibit the cell. The receptive field borders represent the farthest extensions of the direct effects of a light stimulus on a cell's activity. Bars of light of a particular orientation presented outside of the receptive field of a cortical cell, for example, appear to have no effect on the cortical neuron. However, when the orientation selectivity of the cortical cell is then explored with a bar of light presented within the receptive field, the orientation tuning of the cell may be altered somewhat (see Figure 7.12). Not every cortical cell shows this effect, and the alteration in orientation appears to be limited to about 10 degrees. This phenomenon probably relates to the so-called tilt illusion, in which the perceived orientation of a line depends on the ori-entation of surrounding lines and not on the actual orientation of the line (see Figure 7.12). These contextual effects probably play a role in the pro-cess of integrating information from different parts of a complex scene, and perhaps also in visual illusions like the ones illustrated in Figure 8.7.

Not only can peripheral stimuli affect the responsiveness of a cell when its receptive field is stimulated, but attention paid to a visual stimulus can

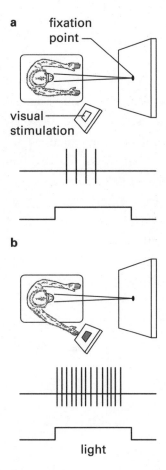

FIGURE 8.10 *The effect of attention on the responsiveness of a cortical neuron. (a) The neuron being recorded is activated by a visual stimulus in the periphery of the visual field. When the monkey is looking at the fixation point in the center of its visual field, the response of the neuron to the peripheral stimulus is minimal. (b) If the monkey attends to the peripheral stimulus by touching it, the response of the neuron is significantly enhanced even though the monkey has not shifted its gaze from the fixation point.*

also alter responsiveness. This was first demonstrated along the *where* pathway in the parietal cortex by Robert Wurtz and Michael Goldberg, working at the National Institutes of Health. If a monkey is looking at a fixation point ahead and a light comes on to its side, but within the receptive field of a neuron being recorded, the cell's response is usually modest. If the monkey is trained to attend to such a peripheral stimulus, by reaching out to touch it when the stimulus comes on, the activity of the neurons being recorded increases significantly (Figure 8.10). Similar effects have been found in the occipital cortex in a number of visual areas, including V_1, V_2, and V_4 (the effect being most pronounced in area V_4, less so in V_2, and even less in V_1). When an animal is paying attention to a particular region of the visual field, the responses to appropriately oriented bars of light presented to receptive fields within the attended part of the visual field are enhanced. The mechanisms underlying these responses and the receptive field alterations in the cortex are not well understood. They may relate simply to long-range excitatory or inhibitory synaptic pathways or to the activation of neuromodulatory influences that alter synaptic strengths through second-messenger cascades or by the action of feedback influences from higher centers. It seems unlikely that they reflect fundamental alterations in cell or synaptic structure because they are reversed rapidly. However, long-term changes can be induced in the cortex by lesions or by prolonged sensory experiences as discussed in Chapter 9.

9

The Dynamic Brain

DEVELOPMENT AND PLASTICITY

The years 1890 and 1891 were periods of intense labour and of most gratifying rewards. Convinced that I had finally found my proper path, I attacked my work with positive fury. During 1890 alone, I published fourteen monographs. My tasks began at nine o'clock in the morning and usually continued until midnight. And the strangest thing is that the work gave me pleasure. It was a delicious rapture, an irresistible enchantment. . . .

I regard as the best of my work at that time the observations devoted to embryonic development of the nervous system, . . . Since the silver chromate [*Golgi method*] yields more instructive and more constant pictures in embryos than in the adult, why should I not explore how the nerve cell develops its form and complexity by degrees, from its germinal phase without processes? . . . In this developmental course, will there not be revealed something like an echo or recapitulation of the dramatic history lived through by the neuron in its millennial progress through the animal series?

With this thought in mind, I took the work in hand, first in chick embryos and later in mammals. I had the satisfaction of discovering the first changes in the neuron, from the timid efforts at the formation of processes, up to the definitive organization of the axon and dendrites. . . . I had the good fortune to behold that fantastic ending of the growing axon. This ending appeared as a concentration of protoplasm of conical form, endowed with amoeboid movements. It could be compared to a living battering-ram, soft and flexible, which advances, pushing aside mechanically the obstacles which it finds in its way, until it reaches the area of its peripheral distribution. This curious terminal club, I christened the *growth cone*.

—Excerpted from Santiago Ramón y Cajal, *Recollections of My Life* (Cambridge, MA: MIT Press, 1989)

One of biology's greatest challenges is to understand how the brain forms. Indeed, how the brain and brain cells develop has long fascinated those studying the brain, as the above excerpt from Santiago Ramón y Cajal's autobiography makes evident. From a few undifferentiated cells in the young embryo, all of the neurons and glial cells that make up the brain arise. The brain consists of hundreds of areas, each carrying out a specific task. Many areas possess neurons unique to those specific parts of the brain. And within each area, the neurons connect with one another, and some project to other areas often considerable distances away. How does all of this happen? What tells an undifferentiated cell to become one kind of a neuron or another? How do axons find their way to distant targets? How do neurons know which cells to synapse on, and whether to form excitatory, inhibitory, or modulatory junctions?

The brain is clearly our most complicated organ, and so far we are ignorant about how its development is orchestrated. In humans, virtually all of our neurons form before birth. If we assume that the brain contains about 86 billion cells, then at least 300,000 neurons are generated per minute in the nine-month gestation period. Even this astonishing figure is an understatement, because cell generation is not constant over the entire period of gestation—as many as 500,000 neurons per minute are formed at times in the first four months!

A long debate has festered over how much environmental factors affect brain development and maturation. How much can brain development be affected by early experience and training? Although virtually all of our neurons are present by birth, and probably no significant increase in numbers of neurons takes place in the brain after about 6 months of age, it takes many years for the brain to mature fully, as described in Chapter 1. How hard-wired is the brain at birth? How malleable is it thereafter? How much of the final product is due to nature, and how much to nurture?

Chapter 1 describes in general terms the development of the mammalian brain, from the formation of the neural tube at the beginning of gestation through to adulthood and beyond. Here I initially focus on the details of how neuronal cells develop and then answer the vital questions posed above. I also turn to the question of how malleable the adult brain may be.

As noted above, as many as half a million cells can be generated per minute in the first four months of gestation in humans. How and when does this happen? Proliferation begins upon closure of the neural tube and initially takes place almost exclusively around its inner surface—an area called the *germinal zone* (Figure 9.1a, b). Initially, the neural tube is just one or a few cell layers thick, but it rapidly thickens, enlarging from the inside out. The proliferation of the neural progenitor cells is under the control both of extrinsic growth factors—specialized proteins that act directly on cells to promote their division—and intrinsic factors: intracellular mechanisms that limit cell division. Cells stop dividing and exit the cell cycle when the negative signals exceed the positive ones, but what the various signals are and how they are controlled are still poorly understood.

Migration and Differentiation of Neurons

When cells exit the cell cycle, they typically move away from the germinal zone and form a distinct layer distally—called the intermediate zone (Figure 9.1b). Cells in the intermediate zone are mainly young neurons that will never divide again. Where they will reside in the brain and even what kind of neuron they are likely to become are now essentially established. Some cells that migrate from the germinal zone do retain the ability to divide, and a number of these cells form important brain structures, including the basal ganglia—subcortical areas that are involved in the initiation of movement. Certain cerebellar cells also proliferate after migration away from the germinal zone, and neural crest cells often divide after they have reached their final destination. In cold-blooded vertebrates, such as frogs or fish, proliferative cells remain in the adult brain and continue to divide and generate new neurons. A particularly clear example is the retina of fish, which continuously adds neurons during the animal's life. In other words, the retina continues to grow as the animal grows over its life span. But most neuroscientists believe this is the exception; in most mammalian species, new neurons are not usually generated in the adult brain. Recent research has, however, identified germinal (stem) cells in at least two regions of the mammalian brain. One is the hippocampus, a region of the brain concerned with the long-term storage of memories. There is evidence that these stem cells in the hippocampus can generate new neurons, but the role of these cells is unclear (see Chapter 11). The other region is the olfactory bulb, but in humans germinal cells may not be present there.

From the intermediate zone, the young neurons must migrate, often considerable distances, to take up their final position. How this happens varies from region to region. In some parts of the brain, such as the retina and spinal cord, cells migrate in response to chemical clues, both positive and negative, present in the area. In other parts of the brain, such as the cortex and cerebellum, specialized glial cells, called *radial glial cells*, provide a scaffolding along which the neurons migrate (Figure 9.1c). The cell bodies of these glial cells reside in the germinal zone, but they extend a branch to the surface of the brain as shown in Figure 9.1b.

Electron microscopy has shown in the intact brain that migrating neurons are entwined around radial glial cell branches, and in tissue culture, neurons have been observed migrating along radial glial cell branches. In a mouse mutant that has a cerebellar defect in which the radial glial cells degenerate early, many of the cerebellar neurons do not end up in their proper positions, and the animals show severe movement deficits. In normal mice (and other animals) the radial glial cells remain until neuronal migration is complete, and then they disappear.

Once the young neurons arrive at their final destination, they are first specified. That is, the kind of neuron they will become is determined. They next undergo differentiation: they extend branches characteristic of the type of neuron they are and begin to make synaptic contacts. What triggers the specification and differentiation of a precursor neuron into a particular cell type? The local environment—the chemical signals the cells encounter—is clearly critical, and this depends on the cells' position in the tissue. In other words, signals from nearby cells determine a cell's fate. Thus, extrinsic signals are key in the process. However, over time, the options for a cell to become a particular type of neuron are limited. That

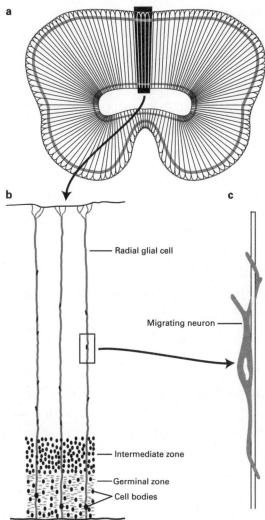

FIGURE 9.1 *The developing cortex. (a) Cells proliferate along the inner surface of the neural tube in the germinal zone. (b) After exiting the cell cycle, they move to the intermediate zone. (c) The cells in the intermediate zone migrate to their correct position in the cortex along radial glial cells. For further details, see text.*

is, a precursor neuron is receptive to a specific inducing signal for only a particular window of time. Thus, intrinsic mechanisms are also at play in neuronal specification and differentiation and are also clearly important. To summarize, to become a particular type of neuron, a precursor cell must be in the right place at the right time during development.

How might environmental and cellular interactions govern cell fate in the developing nervous system? Studies on the developing eye of the fruit fly have provided important insights. The fruit fly eye, like that of the horseshoe crab, has hundreds of photoreceptor units or ommatidia (see Figure 5.2). In the fruit fly, each ommatidium contains eight photoreceptor cells that are precisely arranged in the structure. Each of the photoreceptor cells can be identified, and they are labeled R1 to R8 (R represents "retinular," the technical name for these photoreceptor cells).

During development of the ommatidium, the R8 cell differentiates first, followed by the R2 and R6 cells, which differentiate simultaneously. Then the R3 and R4 cells differentiate, followed by the R1 and R5 cells. The R7 cell, which contains a visual pigment that absorbs in the ultraviolet region of the spectrum, is the last to form.

This strict sequence of development suggests that the earlier cells are responsible for the differentiation of the later cells; indeed, if the developmental sequence is disturbed, the ommatidium does not form properly. A mutation in fruit flies in which the R7 photoreceptor does not form at all—discovered because the flies are insensitive to ultraviolet light—elucidates the nature of the signaling mechanisms and intracellular pathways. The mutant is called *sevenless*. The mutated gene in normal flies codes for a sizable protein that extends across the cell membrane. On the outside of the cell, the protein is like a receptor; on the cell's inside, it has a kinase-like structure. (Recall from Chapter 3 that kinases are enzymes that add phosphate groups to proteins and thus activate or inactivate them.) Presumably an extracellular signal binds to the receptor and activates the intracellular kinase. By phosphorylating intracellular proteins, the kinase initiates a biochemical cascade within the cell, leading to differentiation. If the cascade is not initiated, the cell does not differentiate into a photoreceptor cell. Indeed, what is observed in *sevenless* flies is that the precursor cell destined to become the R7 photoreceptor in a normal eye becomes a nonneural cell in the mutant eye. What can be said about the signal that interacts with the *sevenless* gene protein and the downstream intracellular proteins phosphorylated by it? Progress on both fronts has been made, although the story is not yet complete. A second fruit fly mutation, which also hinders a developing ommatidium to form an R7 cell, has provided clues to the intercellular signal and its origins. This mutant, called *bride of sevenless* or *boss*, affects the R8 cell. The defective gene in *boss* mutants codes for a membrane protein in normal flies. The guess is that the portion of this protein on the outside of the R8 cell activates the *sevenless* receptor protein. Thus, the R7 photoreceptor forms when the R8 photoreceptor cell provides a signal to a precursor cell. Since the R8 signal is a membrane protein,

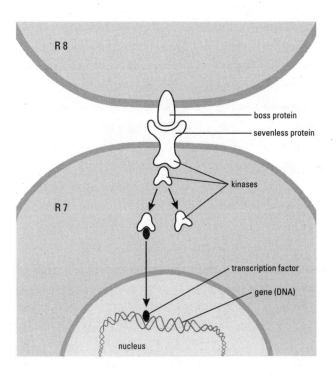

FIGURE 9.2 *A representation of the interaction of R8 with a precursor cell that leads to the formation of the R7 photoreceptor. A protein (boss) on the R8 cell binds to the sevenless protein on the precursor cell, resulting in the activation of the kinase associated with the sevenless protein. Activation of the sevenless kinase leads to the activation of other kinases and ultimately to the activation of transcription factors that regulate gene expression in the nucleus.*

this implies that in ommatidial development direct cell-cell contacts are required to induce differentiation of R7. Figure 9.2 depicts schematically the interaction between R7 and R8.

What about downstream pathways after the receptor has been activated? Several downstream pathway proteins have been identified, and many are kinases themselves. Thus, after the R7 receptor-kinase protein is activated, other kinases are too. Targets of at least one of these kinases are proteins that serve as *transcription factors*. Such factors alter the expression of genes within cells by binding directly to regions of DNA in the nucleus that regulate the turning on or off of genes. The idea is this: when the *sevenless* receptor-kinase protein is activated, appropriate genes are turned on in the precursor cell, which cause it to differentiate into the R7 photoreceptor cell.

In other systems it is likely that diffusible substances control cell differentiation, but the same principles as described above apply. Sometimes the signaling molecules are small proteins called *growth factors*. These proteins activate membrane receptors linked to a cascade of intracellular kinases, ultimately turning on or off specific genes. Thus, the general sequence of events in the developing fruit fly eye is probably true for the differentiation of neurons and glial cells throughout much of the brain.

How Do Axons Find Their Way?

Once neurons begin to differentiate, they extend out branches, both dendrites and axons, as noted in the Cajal excerpt at the beginning of this chapter. This leads eventually to the formation of synapses between neurons and, ultimately, to the wiring of the brain. How do neurons know which cells to synapse upon, and how do the axons of neurons find their way? Sometimes axons must travel substantial distances to their targets.

Again, chemical signaling has long been implicated as playing a critical role in cell-cell recognition. The proposal is that as neurons differentiate they become chemically specified; they make specific proteins on their surfaces that enable other neurons to recognize them.

Early experiments that supported this chemoaffinity hypothesis were performed in the early 1940s by Roger Sperry at the University of Chicago. Sperry studied the projection of retinal ganglion cell axons to the tectum in cold-blooded vertebrates such as fish and frogs. Such projections are quite orderly; ganglion cell axons from one part of the retina project to a particular region of the tectum. Such projections are called *topographic*—they are accurate, consistent, and invariant from one animal to another—hence a retinal map is impressed on the tectum. The right retina projects to the left tectum, and vice versa, and the tectal map is inverted relative to the retinal map. Thus, when a specific region of the retina is activated, a certain region of the tectum responds. Figure 9.3 shows schematically the retinal-tectal projections in goldfish and how the retina maps onto the tectum.

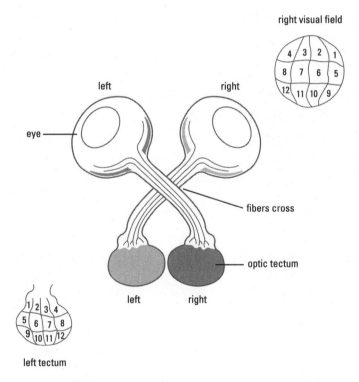

FIGURE 9.3 *Retinal-tectal projections in the goldfish. The ganglion cells from the left retina project to the right tectum and vice versa. The projection is orderly and consistent. Ganglion cell axons from one retinal region end up in a specific part of the tectum, as shown in the maps at the top and bottom. Numbers indicate corresponding areas between the retina and tectum.*

As pointed out in Chapter 6, central nervous system axons regenerate in cold-blooded vertebrates after they have been severed. So, for example, if the optic nerve is cut in a fish or frog, the nerve will regenerate and reform synaptic connections in the tectum. With this reinnervation, vision is restored to the animal. What Sperry did was to cut the optic nerve in a newt (a frog-like animal), rotate the eye 180 degrees in the socket, and then reattach the eye. He observed that after the optic nerves had regenerated, the animal could see but its visual world was upside down and inverted from right to left! When animals so altered were feeding, they consistently misdirected their approaches to food by 180 degrees. If a fly they wished to capture was up and to the right, they moved down and to the left.

These experiments indicated that the severed optic nerve axons had grown back to their original location and vision had been restored. But since the animals' eyes were inverted, they saw an inverted world and responded in this way. Over time, there was no recovery; the animals were permanently altered. The conclusion drawn was that optic nerve axons can recognize the cells they are intended to synapse upon; the cells have complementary markers that allow for a mutual recognition. Many experiments have since been carried out supporting this general idea. However, the bulk of the experiments do not support the notion that a specific retinal axon is wired to a specific tectal cell. Rather, the view today is that axons in cold-blooded animals know in a general sense where they are to go—to which area of the brain and where in that area they should make synapses. But there is some flexibility during both development and regeneration; initial synapses may be broken and new ones formed during development, maturation, and normal functioning of the brain.

Axons grow via specialized structures, called *growth cones*, that are flattened expansions of the tips of growing axons from which extend numerous fine branches (Figure 9.4). Growth cones were first observed by Santiago Ramón y Cajal, and a wonderful and graphic description of these structures is included in the excerpts from his autobiography at the beginning of the chapter. While axons grow, the growth cone is in constant motion, extending and retracting its fine processes and exploring the surrounding area. As the growth cone moves along, it adds new membrane to the axon, and hence it lengthens.

The rate and direction of growth cone movement depend on several factors: the substrate on which the growth cone is moving, the presence of chemicals in the environment, and the electrical fields in the axon's vicinity. Substrate texture and adhesiveness can affect growth cone movement as well as recognition molecules present in the substrate. Diffusible molecules in the environment can also affect growth cone movement—both positively and negatively. Thus, some substances encourage growth cone movement, while others inhibit it. Axons growing long distances can also be guided by specialized cells called *guidepost neurons* found at intermediate distances along the way.

The guess is that the guidepost neurons secrete an attractant chemical sensed by the growth cone. Axons grow toward the guidepost neuron but do not stop when they reach the cell. Rather, when approaching a guidepost neuron, they encounter a chemical that inhibits them and so

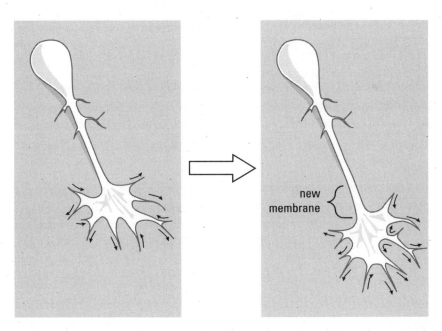

new
membrane

FIGURE 9.4 *Drawings of an axon growing out from a neuron, and its growth cone. The growth cone is continually extending out fine branches that explore the surrounding area. The axon grows by adding new membrane adjacent to the growth cone.*

they move on toward the next guidepost cell. If guidepost cells are damaged, axonal pathfinding may be disrupted.

Initially axon pathfinding in the brain occurs early in development when distances between structures are much shorter than they are in the adult. So another explanation of how axons can find their way over long distances is that early axons serve as *pioneer axons*. Early in development a few axons from any one region find their way to their targets, and as the brain grows, the axons grow as well. Axons from neurons that differentiate later find their way by growing along the pioneer axons.

When axons reach their targets, they then make synaptic connections with appropriate cells—the ones they recognize. Synapse formation often requires reciprocal interaction. Substances released from the growth cone, including neurotransmitters, neuromodulators, and other molecules, initiate postsynaptic structures to form. And the postsynaptic element also releases substances that induce the growth cone to develop into a mature presynaptic terminal.

The Maturing Brain

A most surprising feature of brain development is that many neurons die during the maturation process. In many brain regions, 30–75 percent of the neurons die during development. Why? Much of the cell death appears to relate to a competition for target cells and synaptic sites. Neurons that successfully form synapses with target cells survive, while neurons that lose in the competition for a target cell die. If target cells are eliminated—by removing part of the target, for example—neuronal cell death increases.

Survival of presynaptic neurons depends, therefore, on signals from target cells. Again, these are believed to be chemical signals. One such signal is a protein termed *nerve growth factor (NGF)*, which was extensively studied by Rita Levi-Montalcini, an Italian scientist working at Washington University in St. Louis and in Rome, who discovered the substance. She found that if excess NGF is made available during development, presynaptic cell death diminishes, and if an antibody that inactivates NGF is given to an animal during brain development, neuronal cell death is excessive. She and others have shown that NGF does more than promote survival of neurons. When given to young animals, it can enhance the number and extent of dendrites and multiply the synapses made by those neurons. Moreover, NGF can promote axonal growth. For the discovery of NGF and its significance, Levi-Montalcini was awarded the Nobel Prize in 1986.

Another key feature of the brain maturational process is that once a synapse forms it is not necessarily permanent. Indeed, during brain development axons typically have numerous terminals, and they make more synapses than they do in the fully mature brain. During brain maturation, therefore, axon terminal fields and synaptic contacts are extensively pruned and rearranged to become more restricted, and the pruning is done mainly by glial cells. This means that neurons establish appropriate connections when they develop, but during maturation the connections are rearranged to provide the more precise wiring found in the adult.

The big questions are how long the plasticity of neuronal structure and connectivity lasts, and how it is influenced by experience and environment. For certain regions of the brain, especially ones concerned with higher neural processing, substantial neuronal modification and synaptic plasticity can continue throughout life. We learn and remember new things throughout life, and the mechanisms for memory and learning involve alterations in neuronal and synaptic processes (see Chapter 11).

The extent of neural plasticity is under intensive investigation. What we know unequivocally is that synaptic wiring of the primary visual cortex can change drastically when visual input is altered. Yet these alterations are most readily induced in young individuals. Such experiments on animals, carried out primarily by Torsten Wiesel and David Hubel at the Harvard Medical School, have had a profound impact on our understanding and thinking about brain maturation. Their experiments provided some of the first evidence that substantial remodeling of brain circuitry can occur. Furthermore, their experiments have relevance with regard to phenomena such as language acquisition; the same principles appear to apply.

Visual System Development and Deprivation

How mature is the visual system at birth in a visually inexperienced animal? Are the cells wired up correctly at birth? The answer is simple.

Recordings from newborn monkeys and cats indicate that the responses of neurons in the visual cortex are surprisingly adult-like in their behavior. Many cells are less vigorous in responding to visual stimuli than are neurons in older animals, and a few neurons fail to respond at all to visual stimuli. But neurons with good orientation selectivity are evident, and they may have simple, complex, or specialized *complex receptive field* properties quite like those seen in the adult cat or monkey. The conclusion is that at birth the visual cortex in a cat or monkey is wired up quite correctly and the neurons respond like adult neurons. Thus, visual experience is not necessary for establishing the complex wiring of the cortical neurons and, by inference, also the wiring of retinal and lateral geniculate nucleus (LGN) neurons. The system is ready to go at birth.

If, however, a young monkey or cat is deprived of form vision in one or both eyes, severe alterations in vision can develop. Such a situation can happen in young humans if, for example, the lens of the eye is clouded by a cataract. In monkeys or cats, a similar situation can be induced by sewing an eyelid shut or by applying a light diffuser to the eye. In all of these situations, light can reach the photoreceptors, so the deficit is not the result of light deprivation. Rather, it arises because sharp images do not form on the photoreceptor array; only very fuzzy images reach them.

In humans and animals, visual acuity is severely reduced in such conditions. If the deprivation is monocular, visual acuity measured in the deprived eye is sharply depressed, whereas visual acuity in the other eye is fine. If both eyes are deprived, visual acuity measured in either eye is decreased. This loss of visual acuity is termed *amblyopia*. To understand the changes in the visual system after deprivation, Wiesel and Hubel recorded from neurons in the primary visual cortex of cats and monkeys that had one eyelid closed at birth or shortly thereafter. The recordings involved animals 4 months old or older. Before the experiments were undertaken, the closed eye was opened so the two eyes could be stimulated equivalently. However, virtually all the cells recorded in the cortex received input only from the eye that had been open from birth. The few cells driven by stimulation of the deprived eye usually responded abnormally (Figure 9.5).

When cells in the retina or LGN were recorded, the responses were quite normal. This means that the bulk of the changes were in the cortex and that the structures providing input to the cortex were relatively unaffected by the deprivation.

What is going on in the cortex? The physiological studies described above suggest that input from the open eye occupied a disproportionate amount of the cortex compared to the closed eye. As described in Chapter 7, the cells that receive the bulk of their input from one eye are clustered in columns or stripes, about 0.5 millimeters in thickness, that run across the cortex. The stripes alternate, so one stripe has cells that are primarily driven by the right eye, the next stripe by cells from the left eye, and so forth. In form-deprived animals, this balance is dramatically altered—the open eye occupies territory belonging to the closed eye.

This striking fact can be illustrated anatomically by injecting a

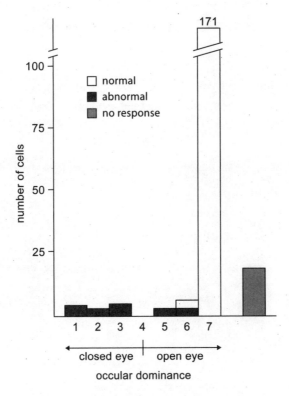

FIGURE 9.5 *Ocular dominance histograms of cell responses recorded in the cortex of a cat that had one eyelid closed from the first to the fourteenth week of life. In the normal animal, relatively few cells have input from just one eye; the bulk of the cells have input from both eyes but to varying degrees (see Figure 7.9). Here virtually all the cells had input from just the open eye.*

radioactive amino acid into one eye and looking at the pattern of radioactivity in the cortex. How this works is as follows. The radioactive amino acid is taken up by ganglion cells in the injected eye. The amino acid is transported along the ganglion cell axons to the LGN. Some of the labeled amino acid is released from the ganglion cell axon terminals, taken up by the LGN neurons, and transported to the cortex by way of their axons. This takes about a week, at which time the terminals of LGN axons that received input from the injected eye are radioactive. A flat section through the cortex, placed on a piece of photographic film, demonstrates the distribution of the labeled terminals, because radioactivity, like light, exposes silver grains in film.

The anatomical images show that the columns from the two eyes in a normal cortex are about equal in width (Figure 9.6a). In a monocularly deprived animal, however, the stripes representing input from the closed eye are much thinner and even discontinuous (Figure 9.6b). How does this happen? One idea is that LGN axons compete for cortical space and synaptic connections in the young animal, and initially the dendrites of geniculate axons overlap considerably. As long as each eye provides equivalent

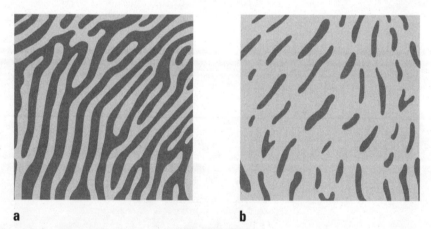

a   b

FIGURE 9.6 (a) A representation of the ocular dominance columns or stripes that extend across the primary visual cortex. The cells in one stripe receive their input from one eye, and the stripes alternate. Each eye has equal representation in the cortex in the normal animal. (b) In an animal in which form vision has been deprived in one eye, the amount of cortex receiving input from that eye is much reduced. The ocular dominance columns or stripes are thin and discontinuous.

input to the cortex, the axons retract equally, and both eyes have equal cortical representation. If one eye has less input or a defective input, the other eye dominates the competition and ends up with more cortical area.

Pattern Deprivation

In the experiments described so far, all form vision was withheld from one or both eyes for a period of time. Is it possible to induce more specific deficits in cortical neurons by restricting one or another aspect of the visual world? One such experiment involves raising animals in environments where they are exposed to bars or stripes of only one orientation. Results from such experiments are consistent: whereas in a normal cortex cells are found that respond to all possible orientations, in animals raised in an environment where they saw only horizontal or vertical stripes the cells responded preferentially to orientations to which they had been exposed (Figure 9.7).

Experiments along the same general idea testing other aspects of cortical neuronal responses have been carried out, with similar results. For example, if animals are restricted at an early age to environments where they see little movement of objects, or movement of objects in only one direction, cells in their cortex appear to be less movement sensitive or are more sensitive to movement in the direction to which they have been exposed.

Critical Periods

Can one induce such severe changes in the visual cortex throughout life? The answer is no. In adult animals (cats and monkeys) and in humans,

vertical stripes

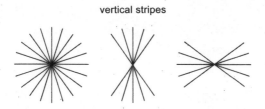

FIGURE 9.7 *Representations of the orientation preferences for cortical cells in a control animal and in animals exposed only to vertical or horizontal stripes. In control animals, all orientations are represented; in animals exposed only to horizontal or vertical stripes, the orientation preferences of the cells reflect the stimuli to which the eyes had been exposed.*

form deprivation does not have dramatic effects on either visual acuity or the responses of cortical neurons. Lid closure or the presence of a cataract for months to years does not cause amblyopia in an adult cat, monkey, or human. Such changes require that the deprivation take place quite early in life. The time when such changes can be induced is called the *critical period*. In monkeys, the critical period is between birth and 1 year of age, with deprivation for the first 6 weeks causing more severe change than deprivation later. In humans, the critical period does not begin until about 6 months of age but extends to 5 or 6 years. Deprivation need not be long during the early part of the critical period for severe changes to happen. A few days of deprivation in the first 2 weeks of a monkey's life can make changes as severe as the ones shown in Figure 9.5.

How easily can the changes be reversed? Surprisingly, they are not readily reversed, even in young animals. If in a young animal a closed eyelid is opened after a short period of deprivation, little recovery is observed after months to years of the eye remaining open. A trick learned long ago by ophthalmologists in treating children who are amblyopic because of crossed eyes can lead to substantial recovery. A patch covering the good eye for a part of each day forces the child to use only the amblyopic eye. Gradual recovery of acuity occurs in the bad eye of the child, and the same occurs with animals rendered amblyopic by monocular form deprivation.

The amblyopia from crossed eyes in children is thus similar to the amblyopia induced by monocular form deprivation. In these children, one eye, usually the straighter eye, gradually becomes dominant, and visual input from the other eye is ignored. The ignored eye becomes highly amblyopic. Such an amblyopia can be induced in animals by altering the eye muscles so their eyes are crossed. Cortical changes in these animals closely resemble the changes after monocular form deprivation. Again, the critical period when changes can be induced matches the form-deprivation critical period.

To summarize, fundamental alterations in the brain can be induced by altering sensory experience or input. The young brain is much more susceptible to these environmental effects than is the more mature brain. Furthermore, some areas of the brain may show striking changes (i.e., visual cortex), whereas other regions (retina and LGN) are minimally changed by alterations in sensory input.

Although what I have said so far suggests that amblyopia caused by form deprivation or crossed eyes cannot be reversed in the adult, several cases are known in which a good eye was lost in an adult and the formerly amblyopic eye recovered reasonable vision. This suggests that recovery may be possible; numerous studies are under way to find the conditions that may permit this, and several of them look promising. Supporting this idea is recent evidence that the cortex does show plasticity—that is, the adult cortex can grow new dendrites and form new synapses.

Plasticity of the Adult Cortex

The notion that the adult brain is quite hard-wired goes back at least a century. The great Spanish neuroanatomist Santiago Ramón y Cajal wrote in 1913 in the conclusion of his work on *Degeneration and Regeneration of the Nervous System*: "In adult centers the nerve paths are something fixed, ended, immutable." However, studies in several cortical areas indicate that significant modifications in cortical structure and function can occur in adults. A number of these relate to changes in response to cortical damage, but others are in response to more normal experiences. Clearly, we can learn and remember new things all our lives, and the cortex is involved in learning and memory, as we shall see. But for many decades this was thought to be a special exception, that most of the adult mammalian brain was "immutable" as Cajal suggested.

Hints that this view is not correct came first, perhaps, from psychological experiments showing that if you place ocular prisms on human beings so that the world they see is upside down, the subjects adapt within a few days and then respond to visual stimuli quite normally thereafter. When the prisms are removed, again the subjects compensate, usually very quickly (in about a day) and they again respond quite normally to visual stimuli.

This result is in stark contrast to experiments on frogs described earlier in which their optic nerves are first severed and then the eyes rotated 180 degrees in the head. In cold-blooded vertebrates, the optic nerve regenerates and the axons grow back to make synapses on the neurons they originally contacted. Following regeneration of the optic nerves, these animals responded exactly as if their visual world is upside down.

The psychological experiments using prisms on human subjects did not teach us anything about the underlying cortical mechanisms involved or even if their compensation was cortical in nature. The first evidence for structural modifications as a result of altered sensory input to the cortex came from studies carried out by Michael Merzenich and his colleagues at the University of California, San Francisco. Using monkeys, they studied how sensory input from the fingers is first processed and represented on the cortex. Somatosensory information, representing touch, pressure, temperature, and pain from all over the body surface, is first processed in the cortex along a cortical strip called the primary somatosensory area, located across from the *primary motor area*. The surface of the body is represented on this area in an orderly and consistent way, although the body representation is not strictly proportional (see Figure 6.7).

By recording from individual neurons in the hand/finger region of the somatosensory cortex and determining which finger is providing the sensory input to a particular neuron (Figure 9.8), Merzenich and his colleagues first found that monkeys vary substantially in how much representation their fingers have on the cortex. Some monkeys have more cortical representation for a particular finger or groups of fingers than others. But of more interest was their finding that if the sensory nerves coming from a finger are cut (called deafferentation), or an entire finger was removed, the representation of the fingers on the cortex changed quite dramatically. Initially, when they recorded from neurons in the area that received input from the lost or deafferentated finger, the neurons were silent (Figure 9.8b). Stimulation of any finger or part of the hand produced no activation in most of the neurons. The exceptions were some neurons on the edges of the area in question, which probably shared some innervation with adjacent fingers.

With time, however, it was possible to activate all the neurons in the deafferentated part of the cortex by stimulating adjacent fingers or, in some cases, other parts of the hand. This took time—weeks, even months—but the adjacent fingers gradually increased their representation and filled in the silent area. The adjacent digits now had a larger representation on the cortex than before (Figure 9.8b). The conclusion from these experiments seems inescapable: new synapses and, presumably, new neuronal branches can be formed in the adult cortex.

A question arising from these experiments is how much reorganization can take place in the adult cortex following deafferentation or loss of a part of the body. In the experiments involving the loss of a finger, the filling in of the silent cortex was relatively limited—it represented alterations in just 1–2 millimeters of cortex. In more extensive deafferentation experiments, carried out in monkeys by other investigators for a different purpose, the innervation to the cortex from an entire limb was cut. Eventually (the recordings were not made until 12 years after the deafferentation) the entire hand-arm region of the somatosensory cortex filled in, a distance of 10–14 millimeters along the cortex.

Merzenich and his colleagues also did converse experiments, looking for cortical changes following extensive manipulation of fingers. These experiments showed that if monkeys were trained to manipulate their fingers to rotate a disk to get food, after several thousand disk rotations,

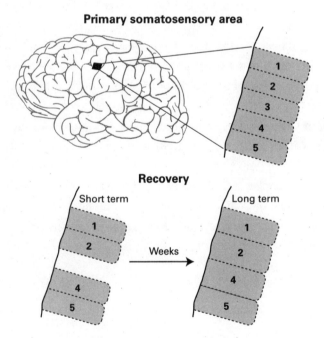

Primary somatosensory area

Recovery

Short term

Long term

Weeks

FIGURE 9.8 (a) Representation of the digits on the primary somatosensory area of the monkey cortex. (b) Reorganization of the cortex following severing of the sensory nerves coming from one finger (digit 3). Initially, the area of the cortex from the deafferentated finger was silent, but with time the area received input from neurons coming from adjacent fingers. The remaining fingers then had an increased representation on the cortex.

over three weeks to several months, the somatosensory cortical area for the monkeys' fingertips had expanded. Furthermore, each cortical neuron whose activity was recorded received input from a smaller area on a particular finger, suggesting a higher touch acuity for these fingertips. If rotation of the disk was limited to just one finger, the cortical expansion was limited to that finger.

It is also the case that the primary motor cortex devoted to finger movement expanded after training. Do similar changes take place in humans who have practiced specialized motor tasks? The most compelling evidence comes from imaging studies on the cortical representation of the left-hand "fingering" and right-hand "bowing" fingers of string (mainly violin) players. The cortical representation for the fingers of the left hand is greater than the cortical representation of the right-hand fingers in string players. As might be expected, the cortices of subjects who learned to play before 12 years of age showed more dramatic increases in left-hand finger representation than those of musicians who began to play later in life. However, subjects who learned to play after age 12 still had a significantly greater representation of the left-hand fingers on the cortex compared to their right-hand fingers. What mechanisms underlie these cortical reorganizations? This is not well understood but, as noted above, is thought to reflect the sprouting of processes and the formation of new synapses within the cortex and to involve neuromodulatory-like cascades as happens in memory and learning and other long-term changes that occur in the cortex (see Chapter 11). The view today is that the adult brain is capable of changing, does show some plasticity throughout life, and that virtually everything we do or experience can cause changes in the brain.

10

Language and Brain Imaging

By the 1920s it was thought that no corner of the earth fit for human habitation had remained unexplored. New Guinea, the world's second largest island, was no exception. The European missionaries, planters, and administrators clung to its coastal lowlands, convinced that no one could live in the treacherous mountain range that ran in a solid line down the middle of the island. But the mountains visible from each coast in fact belonged to two ranges, not one, and between them was a temperate plateau crossed by many fertile valleys. A million Stone Age people lived in those highlands, isolated from the rest of the world for forty thousand years. The veil would not be lifted until gold was discovered in a tributary of one of the main rivers. The ensuing gold rush attracted Michael Leahy, a footloose Australian prospector, who on May 26, 1930, set out to explore the mountains with a fellow prospector and a group of indigenous lowland people hired as carriers. After scaling the heights, Leahy was amazed to see grassy open country on the other side. By nightfall his amazement turned to alarm, because there were points of light in the distance, obvious signs that the valley was populated. After a sleepless night in which Leahy and his party loaded their weapons and assembled a crude bomb, they made their first contact with the highlanders. The astonishment was mutual. Leahy wrote in his diary:

> It was a relief when the [natives] came in sight, the men . . . in front, armed with bows and arrows, the women behind bringing stalks of sugarcane. When he saw the women, Ewunga told me at once that there would be no fight. We waved to them to come on, which they did cautiously, stopping every few yards to look us over. When a few of them finally got up courage to approach, we could see that they were utterly thunderstruck by our appearance. When I took off my hat, those nearest to me backed away in terror. One old chap came forward gingerly with open mouth, and touched me to see if I was real. Then he knelt down, and rubbed his hands over my bare legs, possibly to find if they were painted, and grabbed me around the knees and hugged them, rubbing his bushy head against me. . . . The women and children gradually got courage to approach also, and presently the camp was swarming with the lot of them, all running about and jabbering at once, pointing to . . . everything that was new to them.

The "jabbering" was language—an unfamiliar language, one of eight hundred different ones that would be discovered among the isolated highlanders right up through the 1960s. Leahy's first contact repeated a scene that must have taken place hundreds of times in human history, whenever one people first encountered another. All of them, as far as we know, already had language: Every Hottentot, every Eskimo, every Yanomamo. No mute tribe has ever been discovered, and there is no record that a region has served as a "cradle" of language from which it spread to previously languageless groups.

As in every other case, the language spoken by Leahy's hosts turned out to be no mere jabber but a medium that could

express abstract concepts, invisible entities, and complex trains
of reasoning.

—*Excerpted from Steven Pinker, The Language Instinct*
(New York, NY: Morrow, 1994)

Language is certainly one of the critical features that most distinguishes
humans from animals. There are those who believe that the ability to
speak, read, and write—and thus to communicate ideas and images that
can evoke within us sensations and feelings—is what initiated our rich
inner mental lives that we talk of as awareness or consciousness.

When did language first develop in humans? Human skulls of 150,000
years ago are similar in size to our own, suggesting that these early ances-
tors had brains like ours and were capable of language. However, there
is no evidence for human behaviors that we believe link to language—
rituals and complex social interactions, conceptualization and planning,
art and symbolic representation—until 40,000 years ago. Thus, there is
a gap of about 100,000 years during which we know virtually nothing of
what was going on. Some evidence of modern human behaviors prior to
40,000 years ago has been uncovered—burials, trade, and tool making—
but most paleontologists believe that it was not until 40,000 years ago that
humans were fully modern and that language was universal.

All humans possess language and, as Steven Pinker remarked in his
book, the language instinct : "There are Stone Age societies, but there
is no such thing as a Stone Age language." All human languages are
sophisticated and complex. There are some primitive people who do not
use writing, but all use complex language. It is also true that the ability
to speak is not essential for language; sign languages used by deaf com-
munities can be as sophisticated as spoken languages. Serious attempts
have been made to teach language to certain nonhuman primates, especially chimpanzees. Chimps in the wild can make about thirty-six different sounds, almost as many as English speakers: fifty-two. Each chimp
sound typically conveys something different, whereas each sound we
make (called a phoneme) usually means nothing. We string phonemes
together to make words, and an educated English-speaking adult has a
vocabulary of about 80,000 words.

Is it a difference in vocal tracts and speech abilities that prevents chimps and other nonhuman primates from forming words as we do? One way to test this is to teach chimpanzees sign language, and this has been done, particularly by Duane Rumbaugh and Sue Savage-Rumbaugh at the Yerkes Regional Primate Center in Atlanta. They were able to teach young chimpanzees a vocabulary of about 150 words, but then the animals went no further. These chimps can communicate at about the level of a two-and-a-half-year-old child. However, this is the point at which a child's language abilities are beginning to explode. By age 3 a child typically has a vocabulary of 1,000 words, and by age 4 it might be 4,000 words. Thus, humans are quite distinct from all other animals in their language capability.

Learning Language

Linguists estimate that 6,000 languages are spoken around the world today and thousands more were spoken at one time and are now lost. How can the human brain accommodate so many languages with so much variation? Noam Chomsky, a linguist at the Massachusetts Institute of Technology, studied various languages and noted that there are striking similarities among all of them. He proposed that all languages, present and past, have common grammatical principles. For example, all languages use subjects, verbs, and direct objects. The order in which these elements are positioned in sentences differs among languages, but all languages have these three classes of words. Thus, he suggested that the developing brain possesses innate neural circuits that allow for the acquisition of any of the thousands of languages now or previously spoken.

Certainly all languages share many characteristics, as Chomsky suggested, but whether the brain has innate circuitry to deal with all languages as he proposes or whether it develops at least partially as a result of experience is not certain and is a matter of debate. It is almost certain that both innate mechanisms ("nature") and learned experiences ("nurture") are involved in language acquisition, although the extent to which and how each contributes are not settled. Clearly, innate neural circuitry must place constraints on the ability to make and perceive language, but learning is critical too.

Some of the most compelling evidence that there are innate mechanisms underlying aspects of language—in support of Chomsky's basic idea—is the phenomenon of creolization that occurred in seventeenth-century America. Slave owners brought together people from different African tribes that spoke quite different languages. The slaves quickly created a simplified pidgin language, usually based on the plantation owner's language. Pidgin had a crude word order but lacked a clear grammar. The children of the slaves heard only pidgin but did not adopt it. Rather, they typically created their own languages—creole languages—that had a grammatical structure similar to that of all other human languages.

Another bit of evidence comes from the discovery of a gene defect in a large multigeneration family that has an inherited speech and language disorder. The affected family members have problems with articulating speech sounds, identifying speech sounds, understanding sentences, and with grammatical and other language skills. The gene is inherited dominantly, so about half of the offspring of affected family members have inherited the defect—14 out of 27 offspring in the family initially studied. The gene, called *FOXP-2*, probably codes for a transcription factor that interacts directly with DNA, turning genes on or off. In support of this idea, the protein contains a specific region that is known to bind a target region of DNA. Exactly what the gene does is not yet known, but an obvious suggestion is that it has a role in the development of brain circuitry related to language and speech. The gene appears to have arisen about 200,000 years ago, approximately the time that human brains assumed their modern size and when, it is thought, humans were first capable of some language.

Children appear to learn language in much the same way all over the world. By 1 year of age children begin to speak a few recognizable words, by 18 months they begin to combine words, and by 3 years they can engage in conversation and are speaking in the language or languages to which they have been exposed. Learning a language requires no formal instruction, although hearing it spoken is critical. Indeed, it is thought that even hearing language *in utero* is involved, in that at birth infants prefer the language spoken by their mothers as distinct from

other languages. And clearly, exposure to language early in life accelerates language acquisition and is essential for language development. On
the motor side of language acquisition, babies begin to babble before
18 months, and this also is critical for the development of language.
We have all heard babies babbling "dadada" or "bababa," and this is the
beginning of speech production by infants.

Clearly, young children acquire language much more readily than do
adults, and thus it is generally agreed that there are early critical or sensitive
periods for language acquisition from about 12 months to 6 years. Children
who are exposed to a language in the first 6 years quickly learn to speak
that language perfectly without any detectable accent. After 6 years it is
more difficult to learn a new language, and by puberty the ability to learn
a new language is dramatically reduced. Learning a new language at 40
years of age is similar to learning one at 20, although some people are much
more adept than others at learning new languages. Linguists say that the
accent for a language learned as an adult is never perfect and that a language expert can always tell if someone has learned a language as an adult.
Even many children who learn a new language between the age of 6 and
puberty retain accents characteristic of their native language. The example
often cited is Henry Kissinger, former US secretary of state, who came to
the United States when he was eight years old. He has a distinct German
accent; his brother came at age 6 and is reported to have no accent.

Why do we lose the ability to speak a new language perfectly as we
grow up? Youngsters are sensitive to a broad range of sounds, but they
lose the ability to distinguish or make certain sounds unless they hear
or produce them during the first 6 years or even earlier. For example,
adult Japanese people cannot distinguish an "r" from an "l" sound, yet
seven-month-old Japanese children can distinguish these sounds as readily as American children. By 10 months of age, native Japanese infants
have already lost some of their ability to discriminate "r" from "l" sounds.
American babies, on the other hand, are better at discriminating these
sounds at 10 months than they were 3 months earlier. The conclusion
from these studies is that the period from 6 to 12 months is already critical for babies to learn to discriminate all different language sounds. In
all languages 869 sounds or phonemes have been identified, and infants
6–8 months old can presumably discriminate all of them. After that they

use just a subset—those that they hear and thus distinguish. Conversely, young children can imitate virtually any sound an adult makes, but this ability is also lost with age. By 18 months babies start to make sounds characteristic of the languages to which they are exposed, and their ability to make sounds characteristic of other languages slowly disappears.

What happens if a child is not exposed to any language for the first 6–10 years? Fortunately, there are few recorded cases, but the results are remarkably similar. The most recent example, in the 1960s, is of a young girl, Genie, who at the age of 20 months was locked in a darkened room by her psychotic father. The father and her intimidated brother only growled or barked at her for more than 10 years. When she was 13½ she was discovered and was found to be quite mute. Intensive attempts were undertaken to teach her language, but after 3 years of training she was still unable to speak well; she had the language competence of a 4-year-old at most. The speech she produced was labored and inarticulate. She often was unable to grasp the meaning of speech without contextual clues or gestures, and she was clearly retarded in terms of normal linguistic and comprehensive abilities. Confounding her situation was the fact that she was almost completely isolated during her imprisonment, from both sensory and emotional events, and there was some question as to whether she was mentally retarded. However, Genie's failure to learn language was similar to that of Victor, the "Wild Child of Averyron," who lived alone in the woods in the early part of the nineteenth century. It is conjectured he was abandoned as a young child but managed to survive until he was captured at the age of 12 or 13. Victor, like Genie, never developed normal language skills, despite heroic efforts to teach them to him. There are also cases in the literature of people deaf from very early days having their hearing restored as adults who do not learn to speak effectively.

What is going on in the brain's language areas during the critical period? We cannot, of course, record from neurons in these areas as we can for the visual areas of animals during the critical periods, and so we can only conjecture. It is tempting to suggest, however, that as in the visual cortex during its critical periods, neurons can gain or lose territory, synapses rearrange, and new ones form, depending on language experience. This notion might be extended to suggest that by the age of 6–12 months, neural circuits have formed to discriminate and make all possible language sounds and to acquire grammar. If the circuitry is not used, it is rearranged to accommodate the native language(s) or perhaps even lost. The adage "use it or lose it" might fit for language development as it does for visual development.

Just as with the visual system, different attributes of language acquisition appear to have their own critical periods. The critical period for making sound discriminations might be the earliest; up to 6 or 7 months infants can discriminate all possible human speech sounds, but by 10–12 months this ability is already compromised somewhat and infants might begin to show deficits. With regard to sound production, it appears that up to about 5 or 6 years children can learn to speak a language perfectly, without an accent, although, again, some investigators believe that this critical period extends to puberty, at least for some children. The point to emphasize is that critical periods in language acquisition don't slam shut at a specific age, but there is a gradual decline in various language acquisition abilities over time, superimposed on a considerable variability among people.

With regard to grammar acquisition, a careful study of Korean and Chinese children who came to the United States showed that after 10 years of experience with English as a second language, those who arrived before age 7 had a mastery of English grammar equivalent to that of native English speakers, whereas those who arrived later had grammatical skills that related to their time of arrival in the United States. Of the latter group, those who arrived earlier were more proficient than later arrivals. The grammatical skills of people who arrived in the United States after age 17 were never equivalent to those of earlier arrivals, and it made little difference at what age they arrived. Thus, for grammar the critical period can extend to puberty, but it begins to close as early as 7 years of age, at least for some children.

With regard to learning vocabulary, there appears to be no critical period. We can learn new words, names, and expressions throughout our lives. Certainly, children learn new words faster than adults, and for many people vocabulary learning levels off in high school, but college students show a considerable increase in vocabulary learning, as do graduate and professional school students when they are introduced to the vocabularies of new fields and areas of study. I return to the issue of memory and learning in adults in Chapter 11.

Birdsong

Because language is unique to humans, its development is difficult—indeed, impossible—to study neurobiologically as can be done with the visual system by studying visually inexperienced or visually deprived animals. However, some systems in animals have certain similarities to human language, and these systems can be analyzed in detail. *Birdsong* is one such example.

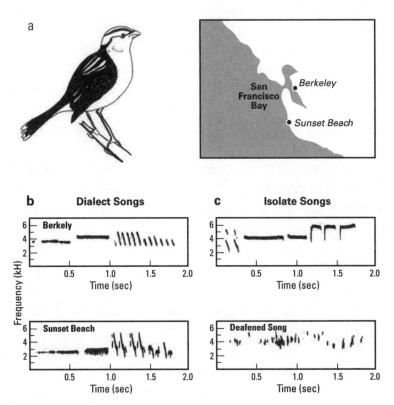

FIGURE 10.1 *White-crowned sparrows look identical in Berkeley and Sunset Beach, two reasonably close locations in the San Francisco area of California (a). However, the songs of white-crowned sparrows in the two locations are distinct (b). These representations of the songs are called sonograms; sound frequency (pitch) is plotted as a function of time. (c) Birds raised in isolation sing, but an isolate song, simpler and distinct from normal songs. Birds deafened before they learn to sing also show very abnormal songs.*

Of the 8,500 species of living birds, about half are songbirds. Birdsong is used for a number of purposes, including attracting mates, defending territories, or simply indicating a bird's location or presence. Birds of the same species have similar songs, but the songs can vary quite significantly over relatively short distances. Figure 10.1b shows, for example, sound spectrograms of white-crowned sparrows recorded in two locations around San Francisco Bay: Berkeley and Sunset Beach. The spectrograms for birds are similar in each of these areas but surprisingly varied between the two areas. Thus, as with human speech dialects, birdsong from different geographical areas varies.

Birdsong, like human language, has great diversity. Some species, like white-crowned sparrows, have one basic song that shows geographic diversity, but other species sing many different songs. Certain wrens, for example, might have as many as 150 songs. The number of song units (syllables) varies considerably with birdsong, from 30 for the canary to roughly 2,000 for a brown thrasher. Again, this is comparable to the variety of speech sounds (phonemes) found in different human languages— which range from as few as 15 to more than 140. Another characteristic of songbirds is that some, such as zebra finches, sing exactly the same song throughout their lives, whereas others, such as canaries, vary their song, incorporating new syllables into their songs from year to year.

How does birdsong develop? Again, we see striking similarities with human language acquisition. Young birds typically learn the songs they hear from their parents. They show a strong preference for songs of their own species, but if exposed only to songs of another species, they can acquire those songs. Indeed, young birds can develop more elaborate songs than their own species sing if they are exposed to such songs appropriately. That young birds have a strong preference for their own species' songs suggests that they have an innate neural circuitry for those songs; however, like humans, they appear capable of acquiring certain other songs as well, so they must also have a circuitry template appropriate to accomplish this. When first learning to sing, birds often exhibit a subsong, noises that might be comparable to babbling in human babies. The young birds next typically produce sounds that contain recognizable bits of the adult song, and finally, they begin to sing the adult song.

Song learning involves two components, song memorization and song

vocalization, and a critical or sensitive period clearly exists for song memorization and perhaps for song vocalization as well. The song memorization
period begins when the birds are about 2 weeks old and lasts for about 8
weeks in white-crowned sparrows and in another model species, the zebra
finch. Although birds can hear before they are 2 weeks old, they do not
memorize their species' song if it is presented to them before the second
week. Conversely, if they do not hear any song until they are 3–4 months
old, they never sing a normal song. Interestingly, birds raised in acoustic
isolation throughout the critical period do eventually sing, but only an "isolate" song that lacks both the spectral and temporal qualities of the normal
song (Figure 10.1c). On the other hand, if a young bird is exposed to the
normal song for only a few days during the critical period, it immediately
acquires the song and sings it accurately as an adult. Young birds can learn
the songs of other species, as noted above, but if the alien song differs substantially from their normal song, the birds develop isolate song singing.
Furthermore, whereas birds can learn alien songs if exposed only to them,
they take much longer to do this than to learn their own species' song.

If, during the beginning of the critical period for song memorization, a
baby zebra finch is exposed to its normal song for 1 week—or long enough
for the bird to memorize the song—subsequent isolation from the normal song or exposure to alien songs makes little difference. The critical
period for song learning is in essence terminated after the bird has been
exposed to the normal song for a week. On the other hand, if the bird is
kept in acoustic isolation for several weeks after the opening of the critical period and then hears the normal song, it learns it rapidly, but this
capacity clearly declines with age and is lost by 3–4 months. How the
song is presented can also be important. Whereas presenting the normal
song with loudspeakers works fine with young song sparrows, it does not
with song sparrows more than 50 days old. The older birds need a live
tutor bird from which to learn the song. In other species, however, a live
tutor is not necessary to train an older bird—a recorded tutor works fine.

Learning to sing by birds—that is, song vocalization—is a distinct
process from song memorization, and song vocalization in white-crowned
sparrows occurs several months after song memorization. Song vocalization might also be constrained by a critical period, although this isn't
entirely certain. Learning to vocalize clearly requires auditory feedback,

so if a bird is deafened by destroying its inner ear after song memorization but before it begins to sing, it does not develop a normal song (Figure 10.1c). If a bird is deafened after it has learned to sing, it usually continues to sing a normal song; acoustic feedback is no longer needed.

Specific areas have been identified in the forebrain of birds that control song production and the learning of song vocalization. There might also be areas specialized for song memorization, but these haven't yet been identified. The song production and vocal learning areas were first identified because of the increased size of certain nuclei (groups of neurons) in male brains at the time that male birds are seeking mates and sing. Two distinct systems have been identified—one in the posterior forebrain that is responsible for song production, and the other in the anterior forebrain that is key for vocal learning. Figure 10.2 illustrates these two systems.

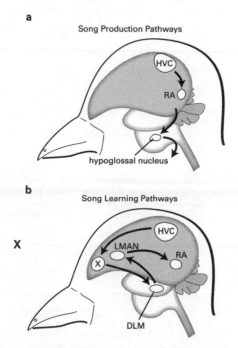

FIGURE 10.2 (a) Neural pathways for birdsong production. When a bird sings, neural activity goes from the higher vocal center (HVC) to the robust nucleus of the arcopallium (RA) and then to the hypoglossal nucleus that contains neurons controlling vocal muscles. (b) When learning to sing, the HVC activates area X, which innervates the dorsolateral nucleus in the thalamus (DLM). Axons from the DLM project to the lateral portion of the magnocellular nucleus of the anterior nidopallium (LMAN), which in turn sends axons to the RA.

Two posterior forebrain nuclei are involved in song production, the higher vocal center (HVC) and the robust nucleus of the arcopallium (RA). When a songbird begins to sing, a wave of neural activity spreads from the HVC to the RA and then to a nucleus in the hindbrain (the hypoglossal nucleus) that contains the motor neurons controlling the vocal muscles. Lesions of the HVC or the RA make birds incapable of producing songs. The anterior forebrain pathway also consists of two nuclei, area X and the lateral portion of the magnocellular nucleus of the anterior nidopallium (LMAN) as well as the dorsolateral thalamic nucleus (DLM). If a lesion is made in the LMAN while a bird is learning to sing, the bird goes no further in song development but is frozen at the level already reached and is incapable of developing a mature song. Lesions in area X also prevent birds from acquiring a stable adult song.

Now that specific groups of cells involved in sound production and vocal learning for birdsong are identified, it will be possible to work out the circuitry of the sound production and vocal learning areas and to uncover how the circuitry is being modified during development.

Language Areas in Humans

Language in humans is controlled mainly by areas in the cerebral cortex, and two areas have been identified as being especially important: Broca's area and Wernicke's area. However, language also depends on our ability to discriminate speech sounds, as well as to make complex speech sounds. Thus, both auditory and motor systems contribute to speech and language, and other neural systems are certainly involved too.

One of the two cortical areas especially important in language, Broca's area, is concerned mainly with the articulation and the production of speech. It is localized in the frontal lobes of the cortex near the region critical for the initiation of face, mouth, and tongue movements in the primary motor cortex (Figure 10.3). Broca's area is named for Pierre Paul Broca, a nineteenth-century French neurologist and anthropologist who studied people who had lost the ability to speak, a condition known as aphasia. He discovered that many of his patients had damage in that part of the cortex that now bears his name. These patients knew what they wanted to say, but their ability to articulate words was impaired. They often could not form proper speech sounds. The first patient Broca studied was called Tan because all he could utter was "Tan, tan, tan" (with an occasional oath thrown in). Lesions in Broca's area also lead to writing deficits and even deficits in sign language, so it is clearly involved in more aspects of language than speech articulation. For example, there is general agreement that Broca's area plays an important role in grammatical processing.

The second language area is called Wernicke's area, named after Carl Wernicke, a German psychiatrist. It is found in the temporal lobe of the cortex, between the so-called primary auditory and visual areas, where sounds and visual stimuli are first processed in the cortex (Figure 10.3). Patients with lesions in this area typically have difficulty with speech comprehension and with reading and writing. They can articulate words perfectly well, but their word choice is inappropriate. The words they utter are clear, but their sentences usually make no sense. As I noted earlier, many neural systems are involved in language, so lesions in other parts of the brain can also cause language deficits. However, Broca's and Wernicke's areas are clearly key for producing meaningful language.

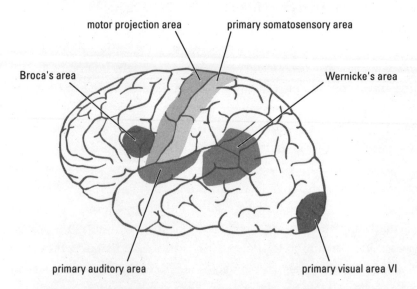

FIGURE 10.3 *Lateral view of the surface of the left side of the cortex. Broca's area is adjacent to the region of the primary motor area concerned with face, tongue, and jaw movements. Wernicke's area is between the primary auditory and visual areas.*

A curious feature of Broca's and Wernicke's areas is that they are most often located on only one side of the brain, typically the left side. In about 95 percent of the population these areas are located in the left hemisphere; in just 5 percent they are located in the right hemisphere. Somewhat more left-handed individuals (about 20 percent) have the language areas on the right side, but even in left-handers, the majority (about 70 percent) have their language areas in the left hemisphere. Interestingly, about 10 percent of the left-handed population have language areas in both hemispheres.

Thus, about 97 percent of right-handers have language areas located on the left side of the brain. What goes on in comparable areas on the brain's right side? Again, clues from the clinic provide insights. Individuals who have lesions in Broca's area and who are highly aphasic often sing very well, which implies that the ability to make music is localized in the right hemisphere, in an area corresponding to Broca's area. Individuals with lesions in the right hemisphere, in the region corresponding to Broca's area, may lose the ability to sing or to play a musical instrument, but they speak perfectly normally.

Why are speech areas confined to just one hemisphere? No one knows. Are there consequences of having the speech areas mainly in the left hemisphere? For normal individuals, the consequences are minimal. Psychologists have shown that subjects will recognize words a bit more accurately and quickly when they are flashed into their right visual fields as compared to the left. (Remember that sensory information on the right side of the body is processed by the left side of the brain.) So something is lost when information is transmitted between the two hemispheres. But the general consensus is that information exchange between the hemispheres, which is mediated by the massive band of axons connecting the two (the *corpus callosum*, depicted in Figure 10.4), is very efficient. In certain individuals, the corpus callosum is cut surgically because of a tumor in the area or incurable epilepsy. In these individuals, clear deficits can be demonstrated. While these individuals can readily name objects placed in the right hand or describe events seen in the right visual field, they cannot verbalize what they are touching or seeing when the object or events are on the left side or in the left visual field. They are aware of the objects or events but cannot describe them.

Their perception of objects and events has been disconnected from their language centers.

Are other faculties localized to just one hemisphere? None that we know of—representation in both hemispheres is the general rule. Yet certain functions seem better done by one hemisphere or the other. For example, in addition to speech, reading, and writing, arithmetic calculations and complex voluntary movements may be mediated primarily in the left hemisphere. For right-handed individuals, skilled movements of the right hand are superior to those of the left hand, and undoubtedly this reflects left cortical dominance. Conversely, in left-handed individuals the right cortex would have dominance.

The right hemisphere in most individuals seems concerned more with

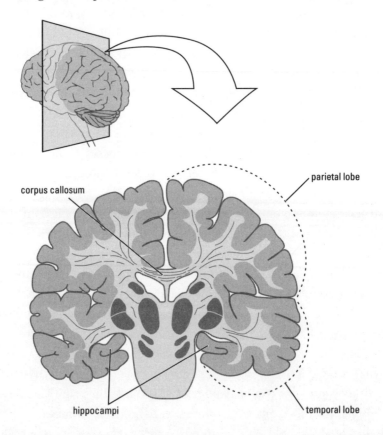

FIGURE 10.4 *A vertical section through the brain showing the corpus callosum, which consists of numerous axons that connect the two hemispheres. It also reveals the hippocampi tucked medially behind the temporal lobes (see Chapter 11)*

complex visual, auditory, and tactile pattern recognition. Spatial sense, intuition, and singing and music making are primarily right-hemisphere functions, as is face recognition in most individuals. Are individuals better in one or another of these attributes because one hemisphere is more developed or dominant? Although this is an intriguing hypothesis, and undoubtedly has some truth to it, concrete evidence in support of the hypothesis is lacking.

Exploring the Human Brain

The human brain is unique in its language ability. It undoubtedly has other unique abilities, not as well defined or understood, that underlie our intelligence and rich mental lives. How can we understand the human brain better? We cannot often record from neurons in human brain tissue or carry out experiments on humans as we do on animals. But the clinic provides a wealth of information, and in this book I give several examples of patients with localized cortical or other brain lesions who have exhibited specific neurological deficits. Broca's area was discovered in this way, as were several other cortical areas. Are there other ways we can gain information about the human brain? One method that has proved useful stems from a surgical procedure to treat *epilepsy*, developed in the late 1930s by Wilder Penfield, a Canadian neurosurgeon. Epileptic seizures occur when a population of neurons in the brain becomes spontaneously active and stimulates abnormal neural activity that sweeps across the cortex. This causes an involuntary and frequently dramatic change in behavior, often accompanied by loss of consciousness. Epilepsy is usually caused by a small group of diseased or damaged cells that initiate the sweep of synchronous brain activity. The purpose of the surgery is to remove the abnormal cells.

The surgical procedure attempts to remove as little brain tissue as possible. To accomplish this, Penfield took advantage of the fact that no pain fibers are in the brain itself; the brain can be touched or even cut and a patient feels nothing, yet neurons can be stimulated electrically from the brain surface. So with a small electrical probe Penfield stimulated discrete areas of the cortex in search of the abnormal cells that lead to a seizure. The brain is exposed by removing the skull overlying the region suspected of harboring the abnormal cells. This part of the procedure is done while the patient's head and skull are anesthetized locally and the patient is asleep. Once the cortex is exposed, the patient is kept fully awake. During the main part of the operation, the surgeon converses with the patient and records sensations, movements, feelings, and even emotions. The surgeon can stimulate the normal cortex in search of abnormal cells and thus has an opportunity to see what effects the stimulation of the cortex can have on a patient. Some of Penfield's results were startling.

Stimulating primary sensory areas evoked sensations appropriate to the area being stimulated. For example, electrical stimulation of the primary visual area in the occipital lobe at the back of the brain usually induced a visual response—a flash or streak of light was reported by the patient. Stimulating the auditory areas on the sides of the brain (temporal lobes) might cause the patient to hear a sound, while stimulation of the primary somatosensory areas in the parietal lobe usually induces the sensation of being touched.

Stimulation of the primary motor cortex (where fine movements are initiated) usually resulted in a small movement of the body—bending a finger or moving part of the face. These effects, which were highly reproducible, prompted Penfield to return to an area previously stimulated and evoke the same response. In doing so, he could map in detail the brain's sensory areas and primary motor area (see Figure 6.7, which is derived from Penfield's maps). A map of the visual field, for example, is on the primary visual area, whereas a representation of the body surface is along the primary somatosensory cortex. These maps are consistent and coherent but distorted. This is clearest in maps of the primary somatosensory and motor areas, where a representation of the body is impressed on the cortical surface. Body areas that have greater touch sensitivity or finer movements—fingers, hand, face, and lips—occupy a disproportionate amount of cortex. More cortex is devoted to their control or sensation than to areas of the body that we control less well or have less sensation, such as our feet and toes. Perhaps the most surprising results obtained by Penfield occurred when he stimulated the more ventral regions of the temporal lobes. On several occasions such stimuli evoked such vivid memories in patients that they believed they were reliving an experience. Not only would they remember an event in exquisite detail, but they felt emotions related to the event. The events were not necessarily major ones in the lives of the individuals. For example, a mother recalled being in her kitchen listening to her small son playing outside. She was aware of neighborhood noises, including the sounds of passing cars.

One of the best-documented examples of temporal lobe stimulation evoking a memory concerned a young woman who heard a song being played by instruments when a specific region of her cortex was stimulated. Whenever that area was stimulated she heard the same song, and the elicited music always started in the same place. The music was so clear to her that she believed it was being played by a phonograph in the operating theater. Later she wrote to Penfield describing what she heard: "There were instruments . . . as though being played by an orchestra. I actually heard it." She added, "I could remember much more of it in the operating room after hearing it than I could three or four days later. The song is not as real to me now as it was in the operating room."

How are such observations interpreted? Because these "experiential" responses were rarely elicited—only about 10 percent of temporal lobe stimulations evoked such responses—firm conclusions have been difficult to make, though it is clear that memories can be evoked by stimulating parts of the temporal cortex. Furthermore, the richness of the evoked memories suggests that memories can be stored in much more detail than we realize or can ordinarily recall. Finally, that patients felt emotions during the evoked memories means that stimulation of the cortex can bring back to consciousness an experience that happened long ago. The fact that such memories are evoked only when regions of the temporal lobe are stimulated suggests that the temporal lobes play a central role in memory storage. Indeed, we now have abundant evidence that a region of the brain tucked behind the temporal lobes, the *hippocampus* (see Figure 10.4), is critical for *long-term memory* consolidation in humans and other mammals.

Brain Imaging

Over the past four decades, powerful noninvasive imaging techniques have been developed that allow study of the brain in awake—even behaving—human subjects. The techniques provide a wealth of new information on human brain structure; in particular, they have shed

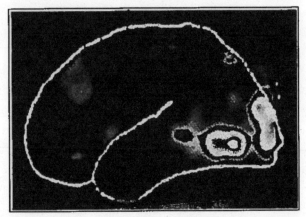

Seeing words

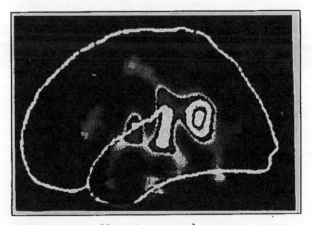

Hearing words

FIGURE 10.5 *PET scanning images from a subject seeing words (top) or hearing words (bottom). When the person sees words, activity increases in the primary visual area and in secondary visual areas along the what pathway. When the person hears words, activity increases in the primary auditory cortex and Wernicke's area.*

light on the functional role of many parts of the cortex, especially the association areas. Brain imaging has revolutionized study of the human brain, enabling neuroscientists to identify the role and function of many brain regions.

The techniques enable investigators to observe which parts of the brain are active when a subject is performing a behavioral task. These methods are based on the fact that, when a brain region is active, blood flow to the region increases. Brain tissue requires oxygen to keep functioning (see Chapter 1); greater blood flow allows for local increases in oxygen and glucose levels necessary for maintaining the neurons. It is these increases in oxygen or glucose that are usually measured. This is accomplished in PET (positron emission tomography) scanning by giving subjects radioactively labeled oxygen or glucose, presenting a stimulus or asking the subject to perform a task, and then determining where hikes in radioactivity occur. The radioactive isotopes are short-lived (and thus safe); the breakdown products of the radioactive atoms are detected and localized by a scanning device surrounding the head. *Functional magnetic resonance imaging (fMRI)*, a more recently developed technique, measures the increase in blood flow directly—it does not require the injection of any substances into the subjects.

The results elegantly extend earlier findings from brain lesions studies or cortical stimulation experiments. Figure 10.5 shows brain scan images of a subject seeing and hearing words. The stimuli or tasks activate distinct areas of the cortex. Activity generated in an area promotes more blood flow, as shown in the images by the white areas. When the subject hears words, activity is greatest in the primary auditory cortex and also in Wernicke's area. When the subject sees words, activity heightens in the primary visual area in the occipital lobes and in secondary or association visual areas along the *what* pathway that extends ventrally in the cortex from the occipital to the temporal lobes.

These data, then, confirm earlier findings on the role of these brain areas, but they can also provide new insights into how the brain processes information or accomplishes a motor task. The power of these imaging techniques is yet to be realized, and their resolution is still crude, yet important new findings are already appearing. A dramatic example was provided by a Danish scientist Per Roland, who wanted to know how motor movements are controlled by the brain. He asked subjects to make a sequence of complex finger movements. He observed, as expected, greater activity in the primary motor cortex in the region where finger movement is initiated. But he also observed activity in another area in the frontal lobes, in one of the *premotor areas* called the *supplementary motor area (Figure 10.6)*. It was known that the premotor areas are related to activity in the primary motor area, but their exact role was unclear. Roland next asked his subjects to rehearse mentally the complex finger movements but not move the fingers. He observed enhanced activity only in the premotor area, not in the primary motor cortex! These data suggest that the supplementary motor area helps in planning and programming motor activity, whereas the primary motor cortex is the locus for initiating skilled motor activity.

At present, many researchers are exploiting these new techniques and improving the methodology. Although it is still too early to draw many firm conclusions, some fascinating and provocative observations have been reported. For instance, Stephen Kosslyn then at Harvard showed that when we form a mental image of something (think of Snoopy, for

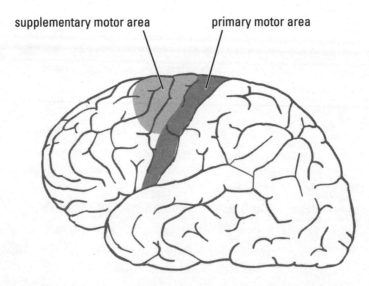

supplementary motor area   primary motor area

FIGURE 10.6 *A surface view of the cortex showing the supplementary motor cortex, a premotor area, adjacent to the primary motor cortex.*

example, with his long droopy ears), the visual areas of the occipital cortex are more active, as though we are actually seeing a picture of Snoopy!

The challenge is to provide a stimulus to a subject or to require a task specific enough that one or a limited number of brain regions show increased activity. It is not easy to present a stimulus or task that allows a crisp interpretation of the result. Although it is simple to activate massive regions of the brain, it is much harder to limit activity to specific brain regions. Another problem is that when subjects are experienced with a task, brain activity patterns can change. Thus, selecting a task that remains novel to the subject upon repetition is essential. These technical issues undoubtedly will be overcome with further experimentation. As resolution improves, so that smaller and smaller brain regions can be visualized, our understanding of the role of brain regions and how they interact while the brain is performing tasks will increase dramatically.

With the enormous success of brain imaging techniques in localizing function to specific brain areas has come the development of other noninvasive techniques to explore the human brain. One of the most elegant of these techniques is *transmagnetic stimulation*. A magnetic beam is focused on a specific brain area that, depending on the parameters chosen, can either stimulate or inactivate the region in question. So, for example, if inactivating parameters are chosen and the beam is focused on Broca's area while the subject is counting backward, the count is interrupted when the beam is on—the subject cannot articulate the appropriate numbers. As soon as the beam is turned off, the subject now articulates the correct number in sequence, leaving the numbers reached when the beam was on unarticulated.

11

Remembering Things
LEARNING AND MEMORY

You have to begin to lose your memory, if only in bits and pieces, to realise that memory is what makes our lives. Life without memory is no life at all. . . . Our memory is our coherence, our reason, our feeling, even our action. Without it, we are nothing . . .

—*Luis Bunuel*

This moving segment from Bunuel's memoir raises fundamental questions—clinical, practical, existential, philosophical: what sort of a life (if any), what sort of a world, what sort of a self, can be preserved in a man who has lost the greater part of his memory and his past. It immediately made me think of a patient of mine in whom these questions are precisely exemplified: Charming, intelligent, Jimmie G., was admitted to our Home for the Aged near New York City in 1975, with a cryptic transfer note saying, "Helpless, demented, confused and disoriented."

Jimmie was a fine-looking man, with a curly bush of grey hair, a healthy forty-nine year-old, cheerful, and friendly.

"Hiya, Doc!" he said. "Nice morning! Do I take this chair here?" He was genial, ready to answer any questions. He told me his name, birth date, and the name of the little town where he was born. He described it in detail, even drew me a map. He spoke of the houses where his family had lived, school days, and his special fondness for mathematics and science. He talked with enthusiasm of his days in the navy— he was seventeen when drafted in 1943. He was a "natural" for radio, and became assistant radio operator on a submarine. He remembered the names of submarines on which he had served, their missions, and even remembered Morse code.

A full early life, remembered vividly, but then his reminiscences stopped. With recalling, Jimmie was full of animation; he did not seem to be speaking of the past but of the present, and I was very struck by the change of tense in his recollections as he passed from his school days to his days in the navy. He had been using the past tense, but now used the present.

A sudden, improbable suspicion seized me. "What year is this, Mr. G.?"

"Forty-five, man." He went on, "We've won the war, FDR's dead, Truman's at the helm. There are great times ahead."

"And you, Jimmie, how old would you be?"

He hesitated a moment, as if engaged in calculation.

"Why, I'm nineteen, Doc. I'll be twenty next birthday."

Looking at the grey-haired man before me, I had an impulse for which I have never forgiven myself—it would have been the height of cruelty had there been any possibility of Jimmie's remembering it.

"Here," I said, and thrust a mirror toward him. "Is that a nineteen-year-old?"

He turned ashen and gripped the sides of the chair. "Jesus "Christ, what's going on?" he whispered. "What's happened to me? Is this a nightmare?"—and he became frantic.

"It's okay, Jimmie," I said soothingly. "It's just a mistake." I took him to the window. He regained his color and started to smile.I stole away, taking the hateful mirror with me.

Two minutes later I re-entered the room. Jimmie was still standing by the window, gazing with pleasure at the kids playing baseball below. He wheeled around as I opened the door, and his face assumed a cheery expression.

"Hiya, Doc!" he said. "Nice morning! You want to talk to me—do I take this chair here?"

"Haven't we met before, Mr. G.?" I asked casually.

"No, I can't say we have. Quite a beard you got there. I wouldn't forget you, Doc!"

—Excerpted from Oliver Sacks, *The Man Who Mistook His Wife for a Hat* (New York, NY: Harper and Row, 1970)

Jimmie G. suffered from a clinical condition called Korsakoff's syndrome. Sergei Korsakoff was a Russian who described patients with this condition in a thesis published in 1887. He wrote, "Memory of recent events is disturbed almost exclusively; recent impressions apparently disappear soonest, whereas impressions of long ago are recalled properly, so that the patients ingenuity, his sharpness of wit, and his resourcefulness remain largely unaffected."

Many of the patients described in the early literature with this disorder had massive and serious brain tumors that were progressive and eventually caused death, but it was also recognized that alcoholism could cause a similar syndrome that was permanent but not progressive. This was Jimmie's problem.

As noted in Chapter 10, Wilder Penfield's experiments stimulating the temporal lobe of the cortex elicited experimental memories, suggesting that the temporal lobes are involved in memory and remembering things. But where in the temporal lobes? There is now substantial evidence that the hippocampus, an area tucked underneath the temporal lobes, is the key structure for being able to remember things (see Figure 10.4).

The crucial role of the hippocampus in the formation of long-term memories was demonstrated dramatically and tragically in 1953 by a young patient who was severely epileptic. This patient (HM) was 27 years old at the time and had frequent and debilitating seizures, so he was unable to work or lead a normal life. It was believed that the hippocampi on both sides of his brain were diseased, and so both were removed surgically. Previously, the hippocampus on one side of the brain had been removed in patients without significant effects. After removing both hippocampi, doctors discovered, much to their dismay, that HM, although cured of his epilepsy, no longer could remember events or facts for more than a few minutes. His *short-term memory* appeared to be unimpaired, but long-term memory mechanisms had been permanently disrupted. Memories of events prior to the operation were retained, but new experiences or facts were quickly forgotten.

HM was studied extensively by psychologists, especially Brenda Milner, a Canadian psychologist, for over 40 years, and virtually no changes occurred in the patient's ability to remember facts or events. He retained such memories for only a few minutes. If he kept thinking about a fact or event, he could continue to recall it for some time; once he was distracted, however, the event or fact was quickly forgotten. Even after 40 years, he did not easily recognize Dr. Milner and she had to reintroduce herself to him whenever they meet. But HM could learn new motor skills or routines and retain them for a long time. (Riding a bicycle is an example of a complicated learned motor skill.) Memories, then, are often classified by psychologists into two types, *declarative memory* and *procedural memory*. The former are memories of facts or events, whereas the latter are retained motor skills or routines. Declarative memories cannot be permanently stored in patients with hippocampal lesions. Procedural memories, by contrast, can be retained in such patients, and there is evidence that the cerebellum participates in the learning and retention of procedural motor skills. When patient HM was asked to perform a motor task to test if he had retained a particular skill, he would deny that he had ever done that task before. His subsequent performance showed clearly that he had retained the skill learned earlier. But even normal individuals are often unaware of the details of a learned skill. For example, when bicyclists are asked what they do when their bicycle begins to lean to the right, most say they would lean to the left. But leaning to the left would increase the right tilt. What is done is to turn the handle bars to compensate for the tilt. Although bilateral surgical excision of the hippocampi is no longer contemplated, there are cases of patients like Jimmy

G. with degenerative diseases involving both hippocampi. These patients are typically unable to form long-term memories, confirming the role of this brain structure in long-term memory mechanisms.

As might be expected, there is enormous interest in the hippocampus among neuroscientists. What is special about this region of the brain? Does it provide clues to how the brain remembers things? The structure of the hippocampus is distinctive from that of other brain regions, but its structure does not provide any special insights into this issue. Yet a striking physiological observation made in 1973 by two scientists, Timothy Bliss and Terje Lomo, who were then working in England on the hippocampus, led to an explosion of research continuing up to the present. Bliss and Lomo discovered that, following strong stimulation of nerve pathways leading into the hippocampus, postsynaptic responses of neurons in the hippocampus were potentiated. That is, the responses of hippocampal neurons to weak stimuli were increased significantly after a strong potentiating stimulus (Figure 11.1). If the strong input stimulus was repeated several times, the potentiation of the responses could be induced to last for days or even weeks. The phenomenon, known as *long-term potentiation (LTP)*, indicates that long-term changes in synaptic efficacy can be induced in single hippocampal neurons by priming stimuli. LTP thus appeared to be a model for how long-term changes can be induced in the brain during memory formation.

Much has been discovered about the mechanisms underlying LTP, but many of the details are still unknown. Many of the same mechanisms described in Chapter 5 that occur in the *Aplysia* nervous system during habituation and sensitization of the gill-withdrawal reflex appear to be involved. Ca^{2+} and cAMP have been implicated, as has protein phosphorylation. There is some evidence for altered release of transmitter from presynaptic terminals, similar to that in *Aplysia*, but also alterations are induced in the postsynaptic neurons. Neuromodulatory mechanisms, including the activation of second-messenger cascades, are implicated, which results in many biochemical changes in the hippocampal neurons during LTP.

In addition to LTP, the hippocampal neurons can generate *long-term depression (LTD)* in which, after priming stimuli are given, the postsynaptic responses of hippocampal neurons are depressed for days to weeks. Mechanisms like the ones involved in LTP might account for LTD, and these results indicate that neuronal activity can be depressed as well as potentiated on a long-term basis. Some have proposed that LTD is more closely related to memory formation than is LTP, but firm evidence is lacking. There is as yet no unequivocal link between either LTP or LTD and memory formation, although by knocking out (by genetic means) one type of kinase activated by Ca_2+, not only is LTP blocked but the ability of a mouse to navigate a maze is also impaired (see below). These mice

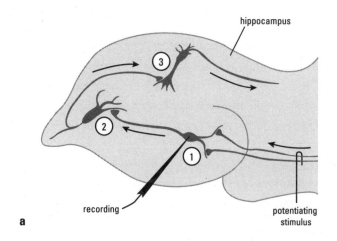

a

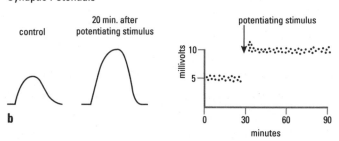

b

FIGURE 11.1 (a) Schematic diagram of the hippocampus. Potentiating stimuli presented to the axons innervating the dentate neurons (1) result in long-term potentiation in these neurons. Long-term potentiation can also be induced in the CA3 (2) and CA1 (3) neurons by providing potentiating stimuli to the axons innervating these neurons. (b) Long-term potentiation. On the left are excitatory synaptic potentials measured before and 20 minutes after a potentiating stimulus. On the right is a plot of synaptic potentials measured before and after a potentiating stimulus.

exhibit a variety of deficits and behavioral changes, so the conclusion that this kinase is critical for LTP and memory storage is still tentative.

The evidence that the hippocampus is essential for long-term memory is unequivocal, but how the hippocampus accomplishes this task is not yet clear. It would appear that long-term memories are not permanently stored in the hippocampus but transferred elsewhere, probably to various regions of the cortex. This idea is prompted mainly by observing patients like HM, who despite having bilateral hippocampal lesions can continue to recall events early in their lives, but not events that happened after their lesions occurred. Exactly how memories are stored in neurons or in neuronal circuits remains a mystery. The best model remains that of long-term habituation and sensitization of the gill-withdrawal reflex in *Aplysia*, in which changes in synaptic number and structure have been observed, suggesting that biochemical and gene expression changes occur in specific neurons during the formation of long-term memories.

A well-established but curious observation is that older long-term memories are more persistent in many forms of brain disease than are more recent memories. With progression of a brain disease, it is not uncommon that a patient's time reference regresses. Such regression happened with my grandfather, who in his 70s and 80s suffered from chronic diabetes. He came to live in the city where I was growing up and initially knew where he was and could correctly identify everyone in our family. As he aged, however, he moved back in time, believing my mother was his sister and I was one of his sons. Furthermore, he thought he was in the Midwest, where he grew up, and not on the East Coast with us. Eventually he identified my mother as his mother and me as a brother. Why older memories are often more resistant to brain disease than are newer memories is not known; when we discover how and where long-term memories are stored in our brains, an answer to this mystery may be forthcoming.

In addition to it being critical for memory consolidation, the hippocampus has two other features that we do not understand very well. First, the hippocampus is the one place in the mammalian brain where it has been established that new neurons are generated in adults. But what their role is remains a mystery. They appear not to live that long, whereas other neurons last a lifetime. It is believed that these new neurons play a role in memory consolidation, but how that happens is not understood. The number of new neurons generated in the hippocampus decreases substantially with age, but environmental factors such as exercise increases their number.

The second noteworthy feature of the hippocampus is that some cells in the hippocampus, termed *place cells*, are activated when an animal enters a specific location in its environment. The population of place cells informs an animal where it is spatially in its environment. Indeed, if you record from a number of place cells in a mouse that is moving around in an environment, you can predict where the animal is. A map of the animal's surroundings is thus encoded by these cells. Interestingly, when an animal moves to a new environment, new place cells appear, and they can be stable for weeks to months. Thus, the hippocampus plays a special role in spatial and temporal recall and representation of environments.

We also now know that cells known as *grid cells* found in a nearby cortical region provide input to the hippocampus and respond whenever an animal is in a regularly spaced grid-like spatial array. Presumably, this information is transmitted; to the place cells in the hippocampus, but exactly the relation between grid cells and place cells is unclear.

Synaptic Mechanisms Underlying LTP

Weak stimuli do not induce LTP by themselves; they must be paired with a strong stimulus—then the weak stimuli show evidence of LTP (Figure 11.1). Neuroscientists talk of this as associative: stimuli must be paired. LTP is associative in another way: to induce it requires activity in both presynaptic and postsynaptic cells; that is, both the cells making the synapses and the cells receiving synaptic input must be activated.

Why both presynaptic and postsynaptic cells must be active to elicit LTP is now understood. It depends on one of the receptor molecules present at synapses on the postsynaptic cell. The presynaptic axon terminal releases a chemical (neurotransmitter) from its synapses when it is activated. The neurotransmitter diffuses across the synaptic cleft to interact with receptor proteins on the postsynaptic side of the synapse. The presynaptic terminal is activated when the voltage across its membrane decreases—the membrane is depolarized.

The neurotransmitter released at hippocampal synapses is glutamate. At synapses where LTP is generated, two types of channel proteins in the postsynaptic membrane interact with glutamate (Figure 11.2). These two channel types are known as *NMDA channels* and non-NMDA channels. NMDA (N-methyl-D-aspartate) is a chemical that specifically activates

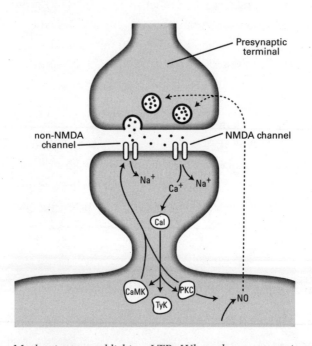

FIGURE 11.2 *Mechanisms establishing LTP. When the presynaptic terminal is depolarized, glutamate is released from synaptic vesicles and interacts with channels on the postsynaptic membrane. The non-NMDA channels allow Na⁺ into the postsynaptic cell, leading to depolarization of the cell and allowing the NMDA channels to be activated and to admit both Na⁺ and Ca²⁺ into the cell. The Ca²⁺ binds to calmodulin (Cal), which in turn activates several kinases (CaMK, PKC, TyK). CaMK and PKC phosphorylate the non-NMDA channels, increasing their effectiveness in admitting Na⁺ into the cell. PKC also promotes the generation of nitric oxide (NO) within the cell, which can diffuse out of the cell and into the presynaptic terminal, thereby increasing its effectiveness in releasing glutamate. Thus, LTP results from enhanced responsiveness of the non-NMDA channels to glutamate and enhanced glutamate release from the presynaptic terminal.*

the NMDA channel; it has no effect on the non-NMDA channels and thus can be used to differentiate the two channel types.

The non-NMDA channels are like most excitatory channel proteins found at various synapses throughout the brain. When activated by glutamate, they immediately open, allowing Na^+ ions to flow into the cell and to depolarize it. However, the NMDA channel works in a more complex manner, and it is key for generating LTP. If the cell is at its normal resting potential glutamate released from the presynaptic terminal binds to the NMDA channel, but its channel opening is blocked. It opens only if the postsynaptic cell is depolarized to some extent. The block is caused by a magnesium ion (Mg^{2+}) ion sitting in the entrance to the NMDA channel at resting membrane voltage.

Depolarization of the cell, which makes the inside of the cell more positive, pushes the positively charged Mg^{2+} ion out of the channel's mouth, and other ions can now enter it. Here again, the NMDA channel is different: whereas most channels allow only monovalent ions that have just one charge (Na^+ or K^+) to flow across the membrane, the NMDA channel allows both monovalent ions and a divalent ion with two charges (Ca^{2+}) to cross the cell membrane. And it is the entry of Ca^{2+} into the cell that is crucial for LTP.

How does Ca^{2+} influx into a neuron lead to LTP? Within the neuron, Ca^{2+} binds to a calcium-binding protein called *calmodulin*. When activated by Ca^{2+}, calmodulin can activate a variety of kinases, our old friends that phosphorylate proteins and thereby alter their properties. Calmodulin can activate at least three different kinases in neurons, but how they increase the postsynaptic response is still not entirely clear. One possibility is that the kinases phosphorylate the non-NMDA channels, thereby increasing their sensitivity to glutamate, or, alternatively, by increasing the amount of Na^+ they permit into the cell following glutamate activation. Phosphorylation of non-NMDA channels is known to do this at various synapses. This, then, is a postsynaptic mechanism.

There is evidence also for an increased release of transmitter from the presynaptic terminal during LTP—a presynaptic mechanism at play. How might this come about? One suggestion for which there is supporting evidence is that kinase activation in the postsynaptic neuron results in the generation of a messenger molecule that diffuses from the postsynaptic cell to the presynaptic terminal and increases synaptic transmitter release. The gas nitric oxide has been implicated as one such messenger molecule, and the enzymes and substrates for the production of nitric oxide are present in many neurons. Figure 11.2 shows the mechanisms for establishing LTP.

Long-Term LTP

I noted earlier that there are short-term and long-term forms of LTP. A single potentiating stimulus produces LTP lasting 1–3 hours, whereas four or more such stimuli produce LTP lasting for days to weeks. Short-term or early LTP can be explained by the mechanisms shown in Figure 11.2, but long-term or late LTP involves more elaborate pathways and more permanent changes in the cells and their synapses. For example, new protein synthesis occurs in long-term LTP but not in short-term LTP.

Figure 11.3 shows schematically the mechanisms involved in long-term LTP. In this process, too, Ca^{2+}-activated calmodulin is involved. If sufficient calmodulin is activated, it interacts with adenylate cyclase, which converts adenosine triphosphate (ATP) to cAMP. The cAMP interacts with protein kinase A (PKA), and this leads to the phosphorylation of the transcription factor CREB (cAMP response element binding protein).

Transcription factors, as described in Chapter 3, interact directly with those regions of genes in the nucleus (the promoter regions) that turn gene expression on or off. When a gene is to be turned on—that is, expressed—the code for the protein to be made is transcribed from the gene's DNA into a piece of another, slightly different nucleic acid, RNA. The *messenger RNA* moves out from the nucleus of the cell to the cytoplasm, where it is translated into protein by structures called ribosomes. In this way, new protein is made that can lead to the strengthening of synapses by, for example, adding new channel proteins to them, or to the formation of entirely new synapses, or even to the development of new branches by the neuron.

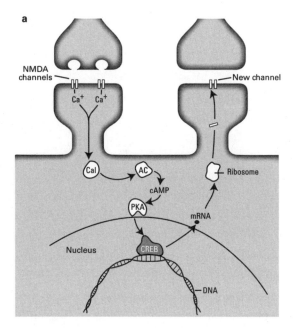

FIGURE 11.3 *Mechanisms underlying long-term or late LTP. Ca^{2+} entering the cell via NMDA channels activates calmodulin (Cal), which in turn activates the enzyme adenylate cyclase (AC). AC catalyzes the production of the second-messenger cAMP, which then activates the kinase PKA. PKA phosphorylates the transcription factor CREB, which interacts with DNA in the cell's nucleus, leading to gene expression and the production of messenger RNA (mRNA). mRNA moves out of the nucleus and interacts with ribosomes, which results in the production of new protein. The newly made proteins can, for example, make new channels that are inserted into the membrane.*

LTP and Memory

None of the experiments described so far provides evidence that LTP is used to store memories or is involved in long-term memory formation. A test of spatial memory, however, has provided impressive evidence linking hippocampal LTP with spatial learning in mice. It is the Morris water maze test, named after R. G. Morris of the University of Edinburgh, Scotland, who developed the procedure. The test works as follows. A mouse placed in a water tank that has spatial cues positioned around its sides quickly learns to find an underwater platform on which it can stand. A trained mouse will rapidly swim to the platform by following the spatial clues, regardless of where in the tank it is released. If the platform is removed, trained mice will go to the area where the platform was and search for it there for at least 60 seconds, the test duration. Once mice learn the test, they can recall where the platform was for weeks to months.

Morris and his colleagues showed that if NMDA receptors were blocked pharmacologically, animals could not learn where the platform was. The mice wandered all around the tank seeking the platform. Additional experiments distinguished short- and long-term LTP effects on spatial learning in mice. Knocking out PKA or the CREB protein yields mice that perform perfectly well in the Morris water maze test for about an hour after training (Figure 11.4). After that they swim randomly around the tank looking for the platform. These mice behave as though they are unable to convert early memories into more persistent memories. Mice given protein synthesis inhibitors show similar deficits in maintaining a long-term memory.

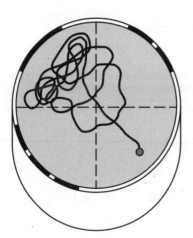 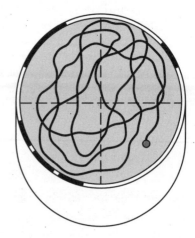

FIGURE 11.4 *The Morris water maze test. A mouse trained to find an underwater platform from visual clues placed around the tank will swim to the quadrant where the platform should be and search intensively for it for the 60-second test period (left). A mouse whose NMDA receptors are blocked or knocked out cannot learn the task, and it will typically wander all around the tank for the duration of the test period (right).*

Short-Term Memory

Individuals who have lost their hippocampi and associated medial temporal lobe structures, like HM, demonstrate perfectly normal short-term memory. When presented with something new, they can remember it initially, and if they "keep it in mind"—focus attention on it—they can retain the memory for some minutes. If distracted, however, they rapidly forget what they had been presented. As discussed above, they have lost the ability to form long-term memories.

Long-term memories, then, depend on the hippocampus, but short-term memory processes must occur someplace other than the medial temporal lobes, because they persist even in individuals without these structures. Furthermore, although both early and late forms of LTP occur in the hippocampus, early LTP does not relate well to short-term memory. It seems more likely that the early and late forms of LTP represent different stages in the formation of long-term memories. In support of this notion, there are differences in the time courses of early LTP and of short-term memory. Early LTP lasts for 1–3 hours, whereas much of short-term memory can be lost in seconds and is entirely gone within minutes at best.

What can we say about short-term memory? We know it is very fragile. A blow to the head easily disrupts it, as does intense mental activity. Individuals in automobile accidents often fail to remember the accident or what happened 10–15 minutes prior to it. It takes time for memories to become stable and resistant to erasure. As noted in Chapter 5, short-term memories may reflect ongoing neuronal activity, whereas persistent long-term memories probably reflect biochemical and structural changes in neurons.

Clearly, much of what we are exposed to on a daily basis we forget—or at least these experiences are not stored as declarative memories that we can readily access. What we retain are those things that are novel, that we focus on and pay attention to, and that have significance or a particular emotional content for us. Most of us wish we had better memories, but individuals who can remember virtually everything—and there are such people—are quite debilitated. They have a difficult time functioning because of all the extraneous material they must constantly deal with. Alexander Luria, a famous Russian neuropsychologist, described an individual like this, D. C. Shereshevskii, in his book *The Mind of a Mnemonist*. Shereshevskii could recount without error long lists of names or numbers presented to him by Luria, but he was overwhelmed in his everyday life by useless information. He was discovered as a mnemonist while working as a reporter because he never wrote down notes for the stories he was writing, even if they involved extensive quotes from people. He recalled the tiniest details but had difficulty understanding the significance of things or generalizing or abstracting them. As a result, his newspaper articles were extensive but unfocused.

Working Memory

Today, many memory researchers describe *working memory* as a form of short-term memory. This is a specialized memory system that holds information for brief periods of time—new information or experiences to which we are exposed, or information we access from memory for planning a particular act or behavior. The information in working memory is rapidly lost—within seconds—from active awareness unless we make an effort to hold onto it, which can be done for several minutes.

The classic example of working memory is looking up a phone number and then remembering it, usually by repeating it mentally, until the call is made. We all know that if we cannot get access to a phone rather quickly we will probably forget the number and have to look it up again. We are also likely to forget the number if we are distracted before dialing.

Working memory has been localized to the prefrontal cortex of the brain, although other brain regions clearly contribute to it. As one might expect, there are substantial connections between the prefrontal cortical areas involved in working memory and the hippocampus, as well as connections with the thalamus, parietal cortex, and Broca's area. Work particularly by Joaquin Fuster at the University of California, Los Angeles and Patricia Goldman-Rakic and her colleagues at Yale University has suggested how neuronal activity participates in working memory. The researchers recorded from neurons in the prefrontal area of the cortex of monkeys while the animals performed a task requiring a delayed response. The monkeys were initially trained to do the task, and then neurons were recorded with metal microelectrodes placed in the appropriate prefrontal cortical area while the monkeys carry out the task (Figure 11.5).

The monkey is first trained to look at a fixation point positioned in the center of its visual field. The animal is further trained to look at this fixation point until it is turned off. While the fixation light is on, a target light appears elsewhere in the visual field for a brief time. The monkey's task is to move its gaze to the location of the target, but not until the fixation light goes off, typically after 5–6 seconds. In other words, the monkey must remember (keep in mind) where the target is for a period of time. If the animal does the task correctly, it is rewarded with a sip of water or grape juice.

It was discovered that some neurons became active when either the fixation or target lights came on, others during the delay period, and still others when the animal began to move its eyes. The neurons of most interest were those that came on during the delay period. They were usually inactive before the task began, while the fixation light was on, and also following completion of the task. However, they were active continuously during the time after the target light was turned off and before the fixation light was turned off. If the delay was made excessively long, the neurons gradually stopped firing and the monkey failed to perform the task correctly at above-chance level. If the monkey was distracted, the neurons also often stopped firing, and again the animal could not do the task reliably. These neurons, then, appeared to be critical for the successful completion of the task—their activity linked the sensory stimuli and the eventual behavior.

The neurons also code for visual field position as well as for the period of delay: some neurons will be active during the delay only when the target appears at the three o'clock position, others when the test target is at the nine o'clock position, and so forth. There is yet a further twist to the story: if the reward is particularly appealing to the monkey—grape juice versus water, for example—the activity of the neurons is generally greater during the delay. This means, of course, that the chances are now higher that the task will be successfully completed, and it suggests how moti-

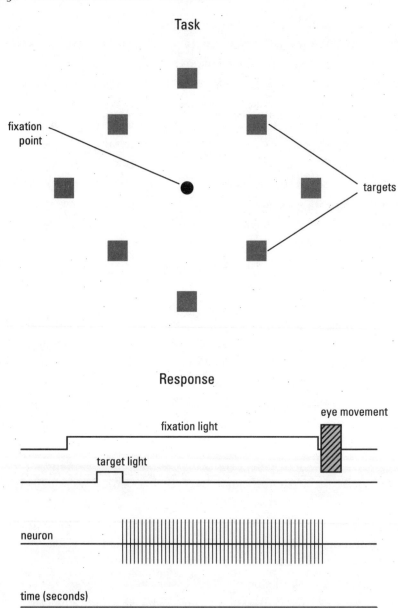

FIGURE 11.5 *Neuronal activity in the prefrontal cortex during working memory. A monkey is trained to look at the fixation light for as long as it remains on. A target light comes on briefly while the monkey is looking at the fixation light. When the fixation light goes off (after 5–6 seconds or so), the monkey must move its eyes to the appropriate target light to receive a reward. Certain neurons in the prefrontal cortex fire vigorously from the time the target light turns off until the eyes move.*

vation can influence the neuronal activity underlying a behavior. These experiments are intriguing because they suggest that information in the working memory system is maintained by ongoing neuronal activity. If this is so, the fragility of short-term memory processes is understandable.

Accuracy of Memories

How accurate are long-term memories that we can recall? The bottom line is not as accurate as we might think or hope. Like visual perception, memories are often reconstructive and creative. What we remember of an event depends on many factors—attention during the event, previous experiences, expectations, biases, and even imagination.

It was long assumed by the law that eyewitness testimony was the best possible evidence—the gold standard—but that view has been seriously challenged. Indeed, two witnesses can give quite different accounts of what they saw. Each recounts what he or she thinks was seen, but not what actually may have happened.

We also know memories can be implanted by suggestion. That is, if people are told on several occasions, or even tell themselves something over and over, they may very well begin to believe it did happen. Autobiographical memories are often a mix of many past memories that are made consistent and coherent by filling in with specific items that may have happened, but of which we do not have a clear recollection. Like visual perception, memory recall can be reconstructive and creative to ensure we live in a coherent world.

12

The Emotional Brain

Elliot was a good husband and father; he had a good job and enviable personal, professional, and social status. Regrettably, his life was soon to fall apart. He first developed severe headaches and found it difficult to concentrate. Subsequently, he appeared to lose his sense of responsibility, and his work often had to be completed or corrected by others. His physician suspected a brain tumor, and the suspicion was correct.

The tumor was located in the lower part of the frontal lobes, approximately in the midline, so that both frontal lobes were being compressed. The tumor was removed, along with surrounding regions of the frontal lobes. The surgery was successful, but following the surgery Elliot was a dramatically different person. He seemed not to have lost any intelligence, but rather, his ability to get things done was severely impaired. He was unable to manage his time appropriately and used poor judgment when trying to accomplish even simple tasks. After repeated incidents and an apparent refusal to take advice and to do things correctly, he lost his job. He tried other jobs but failed at them as well. He squandered his savings on inappropriate

234

ventures, and soon his marriage collapsed. A second marriage also failed. Elliot appeared perfectly healthy and intellectually competent, but he was no longer an effective human being.

Elliot was aware that something was wrong, but when he talked about his life, he did so dispassionately—like an uninvolved spectator. Never did he express sadness, impatience, or even frustration. Seldom did he show any anger—he was constantly calm, relaxed, and detached.

In short, he had lost his feelings—things that once evoked strong emotions no longer did so. He showed very little reaction—either positive or negative—regardless of the situation.

—Adapted from Antonio R. Damasio, *Descartes' Error: Emotion, Reason, and the Human Brain* (New York, NY: Grosset/Putnam, 1994)

I began this book by suggesting that feelings, emotions, consciousness, understanding, and creativity are all aspects of mind. Feelings and emotions—fear, sadness, anger, anxiety, pleasure, hostility, and calmness—localize to certain brain regions. Lesions in these areas can lead to profound changes in a person's emotional behavior and personality, as well as in the ability to manage one's life, as the case of Elliot described above illustrates. As for consciousness, understanding, and creativity, we cannot identify these attributes of mind with a specific brain region. In most textbooks of neuroscience or physiological psychology these aspects of mind are not even brought up, reflecting the fact that little of real substance is known about which brain mechanisms generate them. Yet our fascination with these matters builds as we learn more about the brain.

The aspects of mind that can be explained best are the emotions and emotional behavior, and the consequences of disturbances in emotional behavior. First, as we all know, emotions invariably evoke significant bodily reactions. A frightening experience immediately causes pounding of the heart, rapid breathing, dryness of the mouth, and often sweating. Emotional experiences activate the autonomic nervous system that mediates these bodily effects. It is often difficult to disentangle the emotional reaction to a situation from the bodily effects. Indeed, the well-known Harvard philosopher-psychologist William James suggested at the beginning of the last century that the bodily changes evoke the emotional reactions. When we encounter a frightening situation, the body's rapid response— according to James's theory, pounding of the heart, sweating, dry mouth all lead to the emotional reaction. People who receive less sensory input from the body—patients having a damaged spinal cord, for example—do show more limited emotional responses; these observations appear to support James's theory. Current belief, however, is that emotions result from brain responses to an event.

Conscious emotions are constructed in the brain from signals it receives from both the sense organs and the internal organs. Much as the visual system reconstructs and creates an image from sensory signals (sometimes ambiguous) coming into the visual system as well as from our memories—that is, our experiences and expectations—so do parts of the brain create an emotional reaction from sensory signals impinging on the brain from internal organs and from the outside. And, of course, memories strongly influence an emotional reaction. Indeed, the hippocampus is located just behind one of the key brain structures involved in emotional behaviors (the *amygdala*), and close links between the hippocampus and the amygdala exist. These interactions are reciprocal: we have more vivid memories of events that evoke a strong emotional reaction than of other events. Most of us over 70 can still remember exactly where we were and what we were doing when we heard that President Kennedy had been shot, an event that took place more than 50 years ago!

The Amygdala

The amygdala, a subcortical brain region located just in front of the hippocampus, has been implicated in integrating and coordinating emotional behaviors and in interacting with a number of key brain regions involved in emotional behaviors. (Figures 12.1 and 12.2) The first evidence of this surfaced in the late 1930s when investigators discovered that removing part of a monkey's brain that included the amygdala created a dramatic behavioral change in the animal. Very wild monkeys grew tame after the operation. And they became highly oral, putting virtually everything placed in front of them into their mouths, including repellent things like snakes. The animals also became hypersexual. The key structure underlying this behavioral syndrome is the amygdala, and lesions limited to the amygdala induce tameness and heighten oral and sexual activity in a variety of animals.

When a human's amygdala is electrically stimulated, the subject feels fear and anxiety; stimulating the amygdala of animals leads to a host of autonomic responses and emotional behaviors. The effects of stimulating the amygdala often mimic the effects of stimulating parts of the hypothalamus. Indeed, the amygdala and the hypothalamus interact extensively, and many emotional and behavioral alterations that happen when the amygdala is stimulated are probably mediated by the hypothalamus and the autonomic nervous system.

The amygdala receives direct sensory input from the thalamus, as well as input from areas of the cortex. The direct projection of sensory information from the thalamus to the amygdala is critical for more primitive emotional responses such as fright, whereas information from the cortex probably evokes more subtle and sophisticated emotional responses such

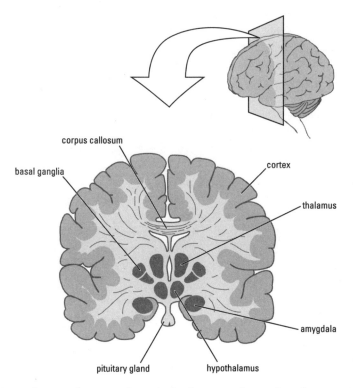

FIGURE 12.1 A vertical section through the brain similar to that shown in Figure 10.4 but a bit more anterior to show the position of the amygdala. The amygdala sits just in front of the hippocampus on the medial side of the temporal lobe.

as anxiety. Having direct input from the thalamus to the amygdala means that the amygdala—and subsequently the hypothalamus and autonomic nervous system—can be activated rapidly by sensory input, which is obviously advantageous in dangerous situations.

The amygdala thus plays a number of roles in emotional responses and behaviors. For example, various emotional reactions such as fear and anxiety require the amygdala. If an animal is conditioned to associate a sound with an unpleasant stimulus such as a shock to the foot, the animal will exhibit anxiety or even fear when it hears the tone in the absence of the stimulus. Lesions in the amygdala will abolish this fear conditioning because they prevent memories of fearful or anxiety-provoking situations from exerting any effects. Anxiety is experienced by most of us at one time or another, but it generally passes fairly quickly. For some, anxiety can be persistent and even debilitating. Extreme anxiety disorders—such as panic disorder, phobias, posttraumatic stress disorder, and obsessive-compulsive disorder—seriously disrupt people's lives.

Persistent anxiety and panic disorder can be treated quite successfully with the benzodiazapines, drugs such as diazepam (Valium) that interact with GABA-activated channels. As described in Chapter 3, GABA is the

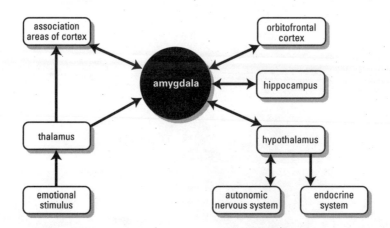

FIGURE 12.2 *The principal inputs to and outputs from the amygdala. Sensory stimuli impinge on the amygdala via the thalamus. The amygdala also receives sensory information from association areas of the cortex. The amygdala interacts reciprocally with the hypothalamus, hippocampus, orbitofrontal cortex, and other cortical areas. The hypothalamus regulates the autonomic nervous system and endocrine system; the expression of emotional effects on these systems is mediated via the hypothalamus.*

primary inhibitory transmitter in the brain; the benzodiazapines bind to the GABA channel, increasing the amount of chloride passing through the channel when it is stimulated. (Barbiturates and ethyl alcohol act similarly on the GABA channels.) In other words, the benzodiazapines enhance inhibition in the brain, and this increased inhibition relieves anxiety. Where do these drugs act principally? Although GABA channels sensitive to the benzodiazepines are found all over the brain, it has been proposed that they reduce anxiety principally by acting in the amygdala. Abundant benzodiazepine-binding sites—GABA channels—are found in the amygdala, providing support for this notion.

But the amygdala also participates in positive emotional reactions. Animals will naturally seek out places or situations that are rewarding— where the animal has enjoyed pleasant experiences of food, water, and sex—and avoid places or situations that are not pleasant or not reinforcing. Again, animals with lesions in their amygdalas do not make these choices.

In addition to interacting with the hypothalamus, the amygdala has strong ties with cortical areas in the frontal lobes, regions essential for the appropriate expression of emotions. One of these cortical regions, just above the orbits of the eye, is called the *orbitofrontal cortex* (Figure 12.3). Lesions in the orbitofrontal cortex dampen normal aggressiveness and emotional responses in animals. Electrical stimulation of this area, producing manifold autonomic responses, indicates that it also interacts

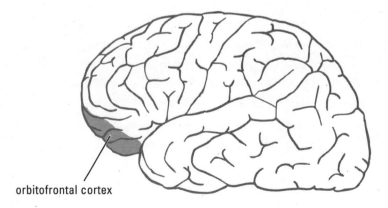

orbitofrontal cortex

FIGURE 12.3 *Surface view of the left hemisphere of the brain, showing the position of the orbitofrontal cortex.*

reciprocally with the amygdala and through the amygdala with the hypothalamus. Figure 12.2 summarizes the inputs, outputs, and interactions the amygdala has with various brain structures.

Hypothalamus

Hypothalamic nuclei regulate heart rate, blood pressure, respiration, gastrointestinal motility, temperature, and so forth, mainly through the autonomic nervous system. The hypothalamus also regulates the release of hormones from the pituitary gland, which sits just below the hypothalamus in the brain (see Figure 6.3). Some of the hypothalamic neurons release small peptides into blood vessels that run between the hypothalamus and the pituitary gland. These blood-borne peptides induce (or inhibit) the release of hormones from cells of the pituitary gland. The pituitary gland hormones enter the general circulation and affect distant cells, tissues, or glands. Hormones released from the pituitary gland affect, for example, the adrenal, mammary, and thyroid glands, the ovaries, and the testes. Control of the endocrine system, especially the reproductive system, is another critical regulatory role of the hypothalamus. During a prolonged emotional or stressful situation, significant changes in endocrine function can result because of the effects of other brain centers on the hypothalamus.

During stress, for example, there is an increased secretion of the small protein corticotrophin-releasing factor from the hypothalamus. This protein causes the release of adrenocorticotropic hormone, a pituitary hormone, which travels via the bloodstream to the adrenal glands that sit just on top of the kidneys. In response to this hormone, the adrenals release several substances, including *cortisol*, a steroid hormone. Over the short term cortisol can help in a stressful situation—it mobilizes energy stores and delivers them to muscles. But if cortisol is continually released in response to a prolonged emotional or stressful situation, harmful results can occur. First of all, cortisol increases appetite, and this leads often to excessive weight gain during stress. But much more serious disorders have also been linked to high levels of cortisol in the bloodstream, including gastric ulcers, colitis, high blood pressure, impotency, and even excessive loss of neurons in the brain. It has even been suggested that prolonged stress can lead to premature aging of the brain and other organs. The immune system is also responsive to cortisol and is depressed when even moderately raised cortisol levels are circulating in the bloodstream. Thus, decreased resistance to disease has been linked to prolonged stress, and in animal experiments an increase in tumor growth has been observed during stress. That students tend to become ill during exam time may relate to the stress they are experiencing. Blood levels of other hormones are also altered by stress or emotional states, and it is likely that harmful effects can result from these changes as well.

The Autonomic Nervous System

The brain oversees two motor systems. One, the *voluntary motor system*, controls the muscles of the limbs, body, and head. It is the voluntary motor system that comes into play when we swing a golf club or swat a fly. The other motor system regulates our internal organs, including the heart, digestive tract, lungs, bladder, and blood vessels; this is the involuntary motor system or *autonomic nervous system* that is regulated by the hypothalamus. As its name implies, control of the internal organs by the autonomic nervous system is mainly involuntary, which is why we are usually unaware of its effects. But this is not always the case—on occasion it is possible to exert voluntary control over internal organs. Furthermore, important interactions occur between the two motor systems. For example, when we voluntarily initiate strenuous activity like running or swimming, more blood flows to muscles critical for sustained activity, and this increased blood flow is mediated by the autonomic nervous system.

Two divisions of the autonomic nervous system exert opposing effects on most organs. One, the *sympathetic nervous system*, is known as the "fight or flight" system. It prepares us for action, and it is rapidly activated when we encounter a frightening or stressful situation. Heart rate and heart output speed up, as does blood flow to the muscles. The pupils of the eye dilate to allow more light into the eye, and at the same time digestive system activity slows down.

The second division, the *parasympathetic nervous system*, is the "rest and digest" system. When it comes into play the body relaxes—heart rate and blood pressure drop, the digestive system becomes more active, and the pupils of the eyes constrict. After Thanksgiving dinner, the effects of the parasympathetic nervous system are obvious: we sink into a soft chair and doze off. Prior to dinner, during the predinner activities, the sympathetic system is in full swing and we are active and energetic.

Animals without a sympathetic nervous system can survive as long as they are maintained in a stable, warm, and comfortable environment. They cannot carry out strenuous activity or survive cold at all well. They also cannot cope with stress. When put in a stressful situation, they may even die, while animals with an intact sympathetic nervous system, when exposed to the same situation, cope perfectly well.

The anatomical organization of our autonomic nervous system, that is, of our sympathetic and parasympathetic nervous systems, is special and distinct. In the sympathetic system, spinal cord neurons extend axons from the spinal cord that synapse upon (innervate) neurons in ganglia that lie along the spinal cord or in the abdominal cavity. The neurons in these sympathetic ganglia extend their axons to the organs they innervate, such as the heart, lungs, and digestive tract. In the parasympathetic system, the ganglia innervating an organ are located in the organ itself. The parasympathetic ganglia are innervated by neurons found in the brain stem or in the lowermost (sacral) part of the spinal cord. Many of the parasympathetic system axons coming from the brain stem are in two cranial nerves (one is the vagus nerve) that exit from the base of the brain and extend branches to various internal organs. The general organization of the autonomic nervous system is shown in Figure 12.4.

The sympathetic and parasympathetic nervous system ganglia are more than relay stations. The synaptic interactions that occur in them are complex and not well understood. Neurons and axons containing neuropeptides and other neuromodulatory substances are located in these ganglia, which means that subtle modulatory interactions likely happen within the ganglia. It is well established, though, that the neurons of the sympathetic ganglia release norepinephrine at their terminals, whereas acetylcholine is released from the terminals of the neurons of the parasympathetic ganglia. Thus, one can regulate the internal organs' levels of sympathetic or parasympathetic activity by giving patients blockers of one or the other of these agents.

How the two autonomic systems regulate an internal organ is elegantly illustrated by their effects on the heart (Figure 12.5). Heart cells have two types of neuromodulatory receptors. One is specific for the norepinephrine released by the terminals of the sympathetic neurons; the other is specific for the acetylcholine released by the parasympathetic

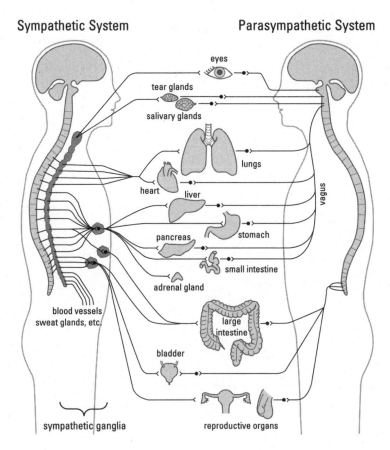

FIGURE 12.4 *Organization of the autonomic nervous system. On the left is the sympathetic system; on the right, the parasympathetic system. In the sympathetic system, spinal cord neurons innervate a series of ganglia that lie next to the spinal cord or three ganglia found in the abdominal cavity. Neurons from these ganglia innervate the internal organs. In the parasympathetic system, motor neurons from the brain stem and sacral region of the brain stem innervate ganglia found in the internal organs themselves. Neurons in these ganglia innervate the host organ. The parasympathetic axons from the brain stem extend to the internal organs in various cranial nerves, particularly the vagus nerve.*

nerve terminals. Both receptors are linked to G-proteins, which in turn are linked to the enzyme adenylate cyclase. However, norepinephrine activates adenylate cyclase via an excitatory G-protein, whereas acetylcholine inhibits adenylate cyclase activity via an inhibitory G-protein. Thus, sympathetic stimulation increases cAMP levels and PKA activity, whereas parasympathetic stimulation inhibits cAMP production and PKA activity.

It turns out that PKA interacts mainly with Ca^{2+} channels in the heart cell membrane, phosphorylating them. Phosphorylation of the channels allows more Ca^{2+} to enter the heart cell; decreased phosphorylation of the channels reduces the amount of Ca^{2+} entering the cells. Ca^{2+} exerts several effects on heart cells. It speeds up the heart rate and makes the

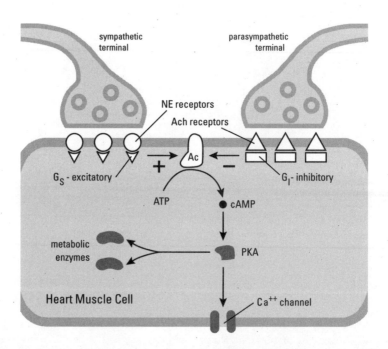

FIGURE 12.5 *Sympathetic and parasympathetic regulation of heart muscle cells. Terminals of sympathetic neurons release norepinephrine (NE), which interacts with receptors linked to an excitatory G-protein (G_S). Parasympathetic terminals release acetylcholine (Ach) which interacts with receptors linked to an inhibitory G-protein (G_I). The G-proteins either stimulate (G_S) (or inhibit (G_I) adenylate cyclase (Ac). In this way, levels of cAMP and PKA are regulated. PKA exerts its effects by altering Ca^{2+} channel activity or enzymatic activity within the cells.*

heart muscle contract more strongly. It also raises the metabolism of the heart cells and induces the heart to beat more strenuously. Thus, the ultimate effect of sympathetic and parasympathetic stimulation on the heart is to regulate intracellular Ca^{2+} levels—increased Ca^{2+} levels stimulate heart activity; decreased Ca^{2+} levels depress heart activity. Thus, we understand regulation of the heart by the autonomic nervous system down to the molecular and ionic level.

Reinforcing Behaviors

Electrically stimulating the hypothalamus in animals evokes virtually every possible autonomic reaction, as was first shown in the 1930s by the American neurophysiologist Stephen Ransom. He also discovered that stimulation of certain hypothalamic regions of animals evoked a panoply of emotional behaviors. Stimulating one hypothalamic region might induce rage or extreme aggressiveness in an animal, whereas stimulating another hypothalamic region might induce placidity in an otherwise wild animal. Stimulation of certain regions of the hypothalamus resulted in what appeared to be a pleasurable experience for the animal. If the experiment was set up so an animal could self-stimulate that hypothalamic region, it would do so continuously, to the exclusion of all other behaviors (Figure 12.6). These regions were originally termed *pleasure centers* and evoked much interest among neuroscientists. Although some brain regions outside of the hypothalamus will induce similar responses— animals will self-stimulate when any of these regions have stimulating electrodes implanted in them—the best and most reliable region to elicit this behavior is in the lateral part of the hypothalamus where a prominent bundle of axons from the mindbrain extends to the forebrain.

Some insight into the significance of the pleasure centers has come from experiments in which antagonists to dopamine were administered to animals, resulting in cessation of the self-stimulation behavior. Earlier experiments had shown that electrical stimulation of the lateral region of the hypothalamus releases dopamine in parts of the brain, particularly in a nucleus located in the basal region of the forebrain called the *nucleus accumbens*, a brain region involved in reinforcing behaviors such as drinking when thirsty, eating when hungry, or satisfying sexual desire. These behaviors have favorable consequences, and so they are repeated, which is why they are called reinforcing or rewarding. During reinforcing behaviors, dopamine is released in the nucleus accumbens, and agents that increase dopamine levels in the brain, including the amphetamines and cocaine, promote these reinforcing behaviors and mimic the effects of electrical stimulation of the lateral hypothalamus. Indeed, laboratory animals will press a lever to self-administer amphetamines or dopamine directly into the nucleus accumbens, much as they will press a lever to electrically self-stimulate the lateral hypothalamus.

The pleasurable and addicting nature of the amphetamines and cocaine are thus believed to be mediated by this reinforcing or rewarding brain system: the release of dopamine in the nucleus accumbens leads somehow to the pleasurable feelings. Other addicting drugs, including the opiates, marijuana, caffeine, and nicotine, also promote the release of dopamine in the nucleus accumbens; again, it seems likely that the pleasurable and addictive effects of these substances

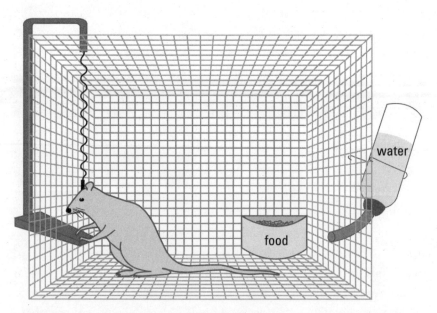

FIGURE 12.6 *Experimental setup for electrical self-stimulation of the brain in the rat. When the lever is depressed, a small electrical pulse is delivered to the brain via an implanted electrode. Animals will do this to the exclusion of eating and drinking.*

relate to this reinforcing or reward system. Even behaviors that are addictive, such as gambling, appear to be related to the release of dopamine in the nucleus accumbens.

Orbitofrontal Cortex

Evidence that the frontal lobes play an important role in emotional behavior can be traced back to the well-known neurological case of a New England railroad construction worker named Phineas Gage. One day in the summer of 1848, Gage was packing dynamite into a hole in a rock with an iron rod when the dynamite exploded prematurely, shooting the rod into his cheek, through the frontal part of his brain, and out the top of his head. No one expected Gage to survive, but miraculously he did. Unfortunately, he changed into a very different man after the accident. Whereas before he was serious, industrious, and energetic, afterward he was irresponsible, irascible, and unpleasant. His self-concern was largely gone, as were his inhibitions. Emotionally he was a different individual whose personality was significantly altered.

John Harlow, the physician who treated him, wrote about the accident and his remarkable recovery, which were reported in Boston and Vermont newspapers. Harlow was well aware of Gage's personality changes and eventually reported them in the medical literature. Five years after Gage died, in 1861 in San Francisco, Harlow arranged to have Gage's skull shipped to Harvard Medical School, where it was kept on display in the Warren Medical Museum. (When I was a medical student there, I remember gazing with awe at the skull and the iron bar that had done the damage. Gage had kept the 13-pound, 3½-foot iron bar, and it had been buried with him. I too was amazed that Gage could have survived such a severe brain injury.) From a modern analysis of Gage's skull we know that the main site of brain damage after the explosion was the orbitofrontal cortex on both sides of the brain. Since then, several other medical cases with damage to the orbitofrontal cortex have been reported. Usually, these patients exhibit severe personality changes and altered emotional responses. The tumor that caused Elliot's problems, described at the beginning of this chapter, caused damage to the orbitofrontal cortex.

Lesions in the orbitofrontal cortex induce a calming effect on aggressive monkeys, and this led to significant medical consequences for humans. A surgical procedure was developed for patients experiencing great emotional distress or extreme anxiety and was performed on thousands of patients, many in mental hospitals. This procedure, the *prefrontal lobotomy*, is discredited today after its widespread use in the 1940s and 1950s proved to be a major mistake. The 1935 experiment that led to the prefrontal lobotomy procedure involved two chimpanzees that would fly into rages whenever they were frustrated. The animals were involved in behavioral experiments, and whenever they made a mistake they displayed violent emotions. Two researchers, John Fulton and C. F. Jacobsen at Yale University, were trying to find out about the role of the frontal lobes in learning and recall. They removed one frontal lobe from the chimps to see what would happen to their performance on a recall task, and this did not change the chimps' behavior. Yet, when they removed the other frontal lobe, things were markedly different. No longer did the animals become frustrated and behave violently; instead they remained placid regardless of whether they made mistakes. They became much easier to handle and seemed much friendlier.

These observations were reported at the 1935 World Congress of Neurology in London. One of the listeners was a Portuguese neurologist, Egas Moniz, who wondered if removing or lesioning part of the frontal lobes in man might not relieve anxiety states and related psychoses. Another paper presented at that Congress reported that when a human patient had his frontal lobes removed, because of a massive tumor, he did not seem impaired intellectually—hence, it was concluded, humans could continue to function well without frontal lobes. Thus, Moniz, working with a neurosurgeon, began to lesion frontal lobes as a treatment for mental illnesses. The procedure did not actually remove the frontal lobes but severed the connections between the orbitofrontal cortex and the rest of the brain. The end result, however, was about the same.

There is no question that the procedure reduced anxiety, obsessions, and compulsions, and very disturbed and violent patients were often rendered more placid and tractable. Yet, many patients had severe side effects, however much they were generally ignored by those promoting the surgery. For example, patients given prefrontal lobotomies often became indifferent to the feelings of others and to the consequences of their own behavior. Whereas pathological emotional behaviors were often eliminated, normal ones were lost as well. The patients also had great difficulty making plans and managing their lives. The operation was too readily performed on inappropriate patients—patients who had a variety of mental disorders unrelated to emotional behavior—and so the surgery often debilitated them even further, which was a disaster.

Today prefrontal lobotomies are seldom performed. Drug therapies are now quite effective in alleviating anxiety states, obsessions, and compulsions and in calming violent individuals. The one malady for which a modified prefrontal lobotomy procedure is sometimes carried out is intractable and excruciating pain. Following the surgery, such patients say that the pain is still present, but it doesn't seem to bother them any longer. The anxiety associated with the pain is relieved.

Rationality

Antonio Damasio, a neurologist formerly at the University of Iowa and now at the University of California Irvine, has focused on the frequently made observation that individuals with frontal lobe lesions such as Phineas Gage or patients who have undergone a prefrontal lobotomy exhibit a dramatic change in personality. They often act irrationally and have difficulty organizing their lives and planning ahead. Some become passive and dependent. And, as I have emphasized, subjects with frontal lobe lesions are disturbed emotionally—they do not respond appropriately to emotional situations, as illustrated by the case of Elliot described by Damasio at the beginning of this chapter. Damasio links the loss of reason in these patients to their debilitated emotional state. He writes, "Certain aspects of the process of emotion and feelings are indispensable for rationality. Feelings point us in the proper direction, take us to the appropriate place in a decision-making space, where we may put the instruments of logic to good use."

Although at first glance this notion appears counterintuitive—we usually think of emotions as interfering with rational behavior—it is also the case, as Damasio points out, that strong feelings incite in us a plan of action. Without emotions and feelings, why bother? And this is the way patients with frontal lobe lesions behave. Elliot is typical of such patients. Not only do they respond passively to emotional situations affecting them, but they also show a lack of concern about others. We think of reason and rationality as among the highest attributes of mind. Their close link to feelings provides a glimpse of the brain creating an aspect of mind.

The Nobel Prize in Economics was given in 2017 to Richard Thaler, who has studied rational behavior and why normal people often behave in nonrational ways economically. That is, they don't often think through the consequences of an action, or lack of action, such as saving for retirement. Once alerted to the value of such an action—that is, attaching emotional value to the action or lack thereof—most people choose more rationally.

An instructive test of these ideas was developed by Damasio, his wife, Hanna Damasio, and one of his collaborators, Antoine Bechara. Subjects are presented four decks of cards from which they must make a selection that results in a reward or a penalty involving play money. The amounts of reward and penalty are different in a complicated way for the four decks: two decks provide higher rewards, but occasionally very high penalties; the other two provide smaller rewards, but almost always small penalties. Since the purpose of the game is to make as much money as possible, and the players are not told how long the game will go on, the trick is to figure out which decks are most advantageous to choose from (namely, those with the lower risk). Normal individuals soon sense that two of the decks (the ones that have the small penalties) are better, and they then choose most often from these decks (Figure 12.7a). When they occasionally do take a chance and choose from the high-risk decks, they show a prominent skin-conductance change—a physiological indicator of an emotional response—and this response builds as the game goes on or until they no longer choose from the bad decks (Figure 12.7b). Changes in skin conductance are the basis for the lie detector test. Individuals who lie usually exhibit anxiety, which promotes sweating via the autonomic nervous system, and skin conductance increases because of the salty sweat released on the skin.

What do patients with frontal lobe lesions do? Although they, too,

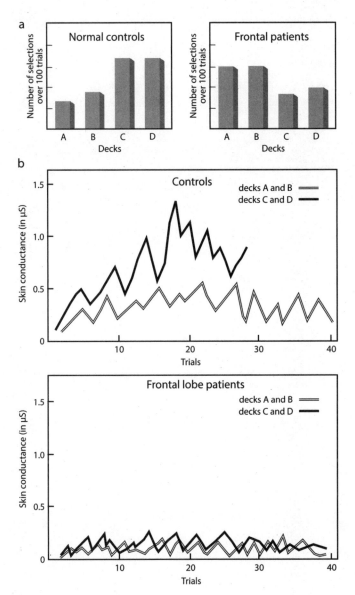

FIGURE 12.7 *Results from the Damasio card game. (a) Normal individuals learn to select cards from the advantageous decks C and D, which provide smaller rewards but smaller penalties, whereas patients with frontal lobe lesions pick from the disadvantageous decks A and B, which have higher rewards but even higher penalties. (b) Skin-conductance changes in normal subjects indicate heightened emotional levels during the game, especially when they are choosing cards from the disadvantageous decks (A and B). Frontal lobe patients exhibit low skin-conductance responses throughout the game, regardless of the decks chosen.*

soon realize which decks are advantageous, they nevertheless tend to pick from the high-risk decks as often or even more often than from the low-risk decks, because the rewards are higher. Often they lose all their money and must borrow money from the investigators to go on with the game. Furthermore, they show no skin-conductance changes throughout the game to any deck choice that they make.

13

Consciousness

Human consciousness is just about the last surviving mystery. A mystery is a phenomenon that people don't know how to think about—yet. There have been other great mysteries: the mystery of the origin of the universe, the mystery of life and reproduction, the mystery of the design to be found in nature, the mysteries of time, space, and gravity. These were not just areas of scientific ignorance, but of utter bafflement and wonder. We do not yet have the final answers to any of the questions of cosmology and particle physics, molecular genetics and evolutionary theory, but we do know how to think about them. The mysteries haven't vanished, but they have been tamed. They no longer overwhelm our efforts to think about the phenomena, because now we know how to tell the misbegotten questions from the right questions, and even if we turn out to be dead wrong about some of the currently accepted answers, we know how to go about looking for better answers.

With consciousness, however, we are still in a terrible muddle. Consciousness stands alone today as a topic that often leaves even the most sophisticated thinkers tongue-tied and confused. And, as with all the earlier mysteries, there are many

who insist—and hope—that there will never be a demystifica-
tion of consciousness.

—Excerpted from Daniel C. Dennett, *Consciousness Explained*
(Boston, MA: Little, Brown, 1992)

No book on the mind can duck the burning question of consciousness.
What is it? What underlies consciousness? We don't know the answers to
these questions, but neuroscientists and others are thinking a great deal
about these questions and attempting to relate consciousness to old and
new findings on brain mechanisms.

We use the phrase "being conscious" quite ambiguously. A person
who is asleep is termed unconscious, as is a patient in a coma. But being
asleep and being in a coma are two very different states. When asleep,
we are often mentally and neurally active, as when we dream. When a
person is comatose, the brain is in a depressed state of activity. Comatose
patients cannot be awakened even by intense sensory stimuli; usually
they are totally unresponsive except for exhibiting basic reflexes such as
the pupillary light reflex.

But even when one is awake and fully responsive, consciousness has
different meanings. For example, much of our everyday activity is carried out "unconsciously." We brush our teeth but usually are not conscious of the activities involved while doing it. We know we brushed
our teeth this morning, but did we need to take the cap off the tooth-
paste tube (or was it already off?), and did we put the cap back on
after we finished? (I just checked to see if I put the cap back on this
morning—I did!)

Awareness is perhaps a better term to use for this kind of consciousness. At any one time we do many things that require mental activity, but
we focus on one activity—what we are aware of at that moment. Later we
can often recall and bring to awareness other activities happening while
we were attending to something else. It is the bringing of something to
consciousness or awareness that we want to understand. Can other mental states shed light on what might be going on when we are conscious
in this restricted sense? Sleep and dreaming have been long thought to
provide clues to mind and consciousness.

Sleep

Why we sleep is still a question we cannot answer. It is clear that we must sleep; indeed, when we are deprived of sleep we become extremely uncomfortable, can hallucinate, and can have psychotic reactions. Depriving animals of sleep can even lead to their death. Clearly, sleeping is one of our strongest drives. We can starve ourselves, or refuse to drink, but it is virtually impossible to stay awake continuously for even a few days.

All mammals sleep, and probably all birds do, too. The evidence for sleep in most cold-blooded animals such as frogs and fishes is more equivocal, but all animals have periods of quiescence that may serve as sleep. It is also true, though, that mammals and birds have distinct stages of sleep that are not seen in cold-blooded vertebrates, so the equivalence of sleep behaviors between various kinds of animals is not clear.

An obvious reason for sleep is that it rests the brain and provides time for restoring neural processes. A corollary is that brain activity is diminished during sleep. But recordings from neurons during sleep indicate that they can be very active then. So sleep is now viewed as an active behavioral state, not an inactive one. It is true that some neurons and certain neural circuits are quiescent during sleep, but many neurons are as active during sleep as at other times, or possibly even more active. One current theory links sleep with long-term memory consolidation, and there is evidence that memories of certain tasks are improved following a period of sleep.

Distinct stages of sleep in mammals have been identified. Upon retiring for the night, we fall into a deep phase of sleep within 30–45 minutes. During deep sleep, parasympathetic nervous system activity predominates; heart rate and blood pressure decline, but gastrointestinal motility increases. Muscles are relaxed, though we move every 5–20 minutes on average. After we have been asleep for about 90 minutes (or in deep sleep for about an hour), sleeping behavior changes. Sleep is shallower, and subjects can be much more easily roused by meaningful stimuli—calling their name, for instance. At the same time, muscle tone slackens and the limbs exhibit a partial paralysis. Even so, the eyes begin to move back and forth, sometimes very rapidly, as do the internal ear muscles. This stage of sleep, termed *rapid eye movement (REM) sleep*, is a highly active phase of sleep. Cerebral blood flow rises, as does brain oxygen consumption. And it is during REM sleep that most dreaming occurs and when some memory consolidation may happen. When subjects are awakened from REM sleep, most report that they have been dreaming; only a minority of subjects awakened from deep sleep say they have been dreaming.

REM sleep initially lasts for only 20 minutes, and then deep sleep ensues again for about 90 minutes. REM sleep recurs—for a longer period of time—and is followed by another phase of a deeper level of sleep. As the night wears on, the levels of deep sleep become shallower and shallower, and REM sleep periods typically become longer and longer (Figure 13.1). Thus, over the course of the night, some four or five periods of REM sleep occur, and each active period is accompanied by dreaming. About one-quarter of the night is spent in REM sleep, more of which is in the second half of the night.

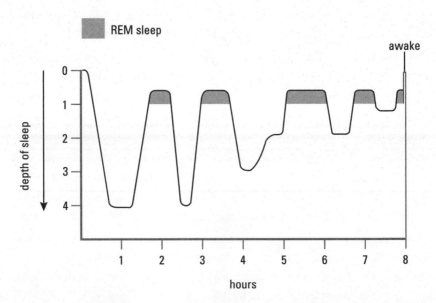

FIGURE 13.1 *The stages of sleep. Within 30–45 minutes of falling asleep, a deep phase of sleep occurs. This is interrupted after about an hour by a shallower phase of sleep and rapid eye movement (REM) sleep. As the night progresses, the periods of deep sleep become progressively shorter and shallower, whereas REM sleep becomes more prominent.*

Dreaming

The recognition of REM sleep and its correlation with dreaming has led to the realization that dreaming is a much more common phenomenon than was previously thought. Since ancient times dreaming has been thought important for understanding mental processes. Indeed, a famous book on dreams, written by Artemidorus of Daldis, dates to the second century A.D., and the dreams recounted in that volume are remarkably similar to modern ones. Yet it was generally assumed until recently that dreaming was a rare occurrence during the night. This is not so; we dream regularly during REM sleep, which means that each night we have many dreams, but most are forgotten within 8–10 minutes or so after the cessation of REM sleep. In other words, most dreams are not consolidated. We ordinarily remember dreams only from the late phases of sleep, from the time just before we wake up. Despite this, when subjects are awakened during REM sleep at any time during the night, at least three-quarters of the time they report that they had been dreaming.

Dreaming is clearly a distinctive state of mind. Most dreams—perhaps as many as two-thirds of them—are unpleasant and associated with apprehension, sadness, or even anger. Only a few are pleasant or exciting. As we all can attest, we are often frustrated in our dreams and subject to hostile acts. Dreams tend to be mainly visual in sighted people, but primarily auditory in blind individuals. Individuals who lose their sight gradually lose the ability to see in their dreams. Dreaming occurs also in deep sleep but seems to be less common. Most nightmares that awaken children and adults in terror probably occur during deep sleep.

What role do dreams play? Much has been said about them, but little has been concluded. Are they unique to humans? All mammals and birds demonstrate REM sleep, and cats and dogs move during REM sleep as though they are dreaming. But whether they dream as we do is not known, just as it is not known whether their levels of consciousness or awareness are similar to ours.

Control of Sleep and Arousal

Lesions in the reticular formation of the medulla can make an animal stuporous or comatose. A comatose animal is totally unresponsive to external stimuli, whereas an animal in a stupor responds only to intense stimuli. As noted earlier, certain neurons in the reticular formation extend their axons widely throughout the brain, including the cerebral cortex, and are critical for maintaining the brain in an aroused state. Three principal neuroactive substances participate in this arousal system: norepinephrine, acetylcholine, and serotonin. Lesions of the reticular formation that destroy these aminergic and cholinergic neurons render an animal permanently comatose.

Neurons in the reticular formation also control sleep in an active fashion; that is, sleep occurs only when these neurons are activated. Furthermore, deep sleep and REM sleep are controlled separately. Certain neurons containing acetylcholine and located in the brain stem facilitate REM sleep. On the other hand, neurons in the brain stem containing norepinephrine and serotonin inhibit REM sleep.

How sleep is initiated is still not well understood. For many years it has been supposed that neurons in the brain release sleep-promoting substances. The search for such agents has been intense, and substances have been isolated, including two small peptides that will promote sleep in animals. But whether any of the substances so far isolated is a key sleep-promoting factor is not clear. Nevertheless, the enormous interest in discovering how sleep is initiated, controlled, and terminated, and by what factors, is undiminished, because insomnia is such a common and disruptive phenomenon. About 15 percent of people in industrialized countries have serious or chronic sleep problems, and another 20 percent have occasional insomnia. Furthermore, insomnia becomes more common among elderly people.

Sleep, like many biological processes, is partially regulated by a biological clock (circadian rhythm; see Chapter 5). Our internal clock, with a periodicity of about 24 hours, contributes to the regulation of sleep. The natural circadian rhythm that most of us possess is a little longer than 24 hours, but light each day resets the clock to approximately 24 hours.

When humans (or other animals) are kept in constant darkness, the clock becomes free-running, and for most animals and humans the circadian rhythm lengthens. This means that sleep comes later and later each day, as does awakening, so that the rhythm becomes out of synchrony with ordinary day and night. Our body's main clock is located in a nucleus found in the hypothalamus. As is expected, this hypothalamic nucleus receives direct input from the retina, which is the pathway by which light resets the clock.

Consciousness and Awareness

Can we begin to classify states of consciousness or levels of consciousness? Antonio Damasio in his book *A Feeling of What Happens* suggests a useful scheme. The first level he calls "wakefulness," which all animals exhibit. That is, they are alert, respond to sensory stimuli, and show appropriate behaviors. They can detect objects and pay attention to them. Fish and frogs clearly exhibit this stage of consciousness.

The next level he calls "core consciousness." Animals that are at this level demonstrate procedural memories—they can be readily trained—and they remember immediate past events. They show emotional responses and can form relationships. Examples are dogs and cats. The final level is called "extended consciousness" and is demonstrated by primates. There is recognition of past events and anticipation of future events, as well as a sense of self. Humans go beyond this and possess language and demonstrate creativity.

What can we say about the neurobiology of consciousness? Throughout this book I discuss how many brain phenomena can be linked with specific brain regions. The control of basic drives and acts occurs in the medulla and hypothalamus. Motor control and sensory processing can be related to a hierarchical series of cortical areas. Initial sensory processing occurs in the primary sensory areas and extends to association areas concerned with more specific aspects of sensory processing. The visual system, for example, has areas that primarily process objects, color, or motion, and some specialized areas, such as those that recognize faces.

Language can also be localized to certain cortical areas, as can different forms of memory. Even phenomena usually associated more with mind than brain, such as emotions and feelings, can be linked to specific brain structures, pathways, and regions. But what about consciousness? Whereas we can render an animal or human unconscious by lesioning the reticular formation in the brain stem, no one has provided any evidence that a region of the brain specifically relates to consciousness or awareness in the sense I have been defining it. This is not to say there is no such area; what I wish to suggest is that there may not be just one. Rather, consciousness may depend on many areas, working in concert.

Clearly, our rich mental life depends on higher cortical function. The rapid development of the cortex during mammalian evolution made a big difference for humans, whose cortex is undoubtedly more developed than that of any other vertebrate or mammal. We possess mental abilities beyond those of any other organism, with language being the most obvious example. Humans can communicate ideas and images to one another either orally or by the written word, and this communication readily evokes images and emotions. We do not need to actually see something to visualize it—a poem or a narrative passage in a book can evoke intense images, sensations, and feelings.

Contrast this behavior with that of a frog, which "sees" quite well; it can visually identify a passing fly such that its tongue captures the fly when it is in front of the frog. But the fly must be moving for the frog to see it. A frog in a cage stocked with perfectly edible dead flies will starve to death; it simply doesn't see them if they are not moving. What, then, does the frog really see? Are its visual responses purely reflexive, much like our knee-jerk reflex? My guess is that they are: an animal without a cortex does not form visual images and does not see in the sense humans do. It responds to visual stimuli, but I would say unconsciously and without awareness. The same situation holds in human patients who have massive lesions in area V1, the primary visual cortex. Such patients have no awareness of seeing, yet they will exhibit visual reflexes. They may blink or even duck when a threatening object rapidly approaches them, but when asked why they blinked or ducked, they cannot explain why. The story of Dee in Chapter 8 illustrates this point very well. She could not recognize objects or faces but had surprisingly sophisticated visual abilities.

Our visual system constructs images based not only on incoming sensory information but also on experience. What we see depends very much on our expectations—what we have seen before—as well as on the information coming from our retinas. As noted in Chapter 8 visual perception is reconstructive and creative. The visual information coming from our eyes is imperfect; we construct a logical image based on incoming information and visual memories. If the visual information arriving by way of the optic nerve is ambiguous, we make a logical percept. But percepts can change, as the famous face-vase illusion beautifully illustrates (see Figure 8.9). To make the percept change requires attention, much as focusing our awareness on something requires attention.

In fact, what we see depends very much on memory and learning. Indeed, we must learn to see. Many documented cases (including one described in Oliver Sacks's book *An Anthropologist on Mars*) tell of individuals who had been blind from birth or since early childhood and whose vision was restored. The cataract or other impediment preventing their eyes from receiving appropriate visual stimuli was corrected. Yet, these individuals never learn to see, even though their "eyesight" is restored. They cannot effectively use visual information. They report seeing colors and vague objects, but they almost never become skilled visually. They continue to rely on other senses to function.

Learning to see occurs in the young human as does learning to speak. Just as it becomes more difficult to learn a language as one gets to be a teenager or older, it is difficult to learn to see when one is already an adult. And it is likely that this principle holds for other sensory modalities.

Consciousness, then, may be a logical extension of the phenomena we have been examining: sensory perception, language, memory, and learning. When we focus our consciousness on something, we form mental images much as we do when we form visual images. We draw on memories to do this but use the appropriate sensory systems to construct the images. Memories and images can evoke emotions and even bodily reactions by way of the autonomic nervous system. Language is a powerful stimulus for evoking such internal images. It may have been what first initiated our rich inner mental lives of consciousness. But now consciousness or awareness can be elicited entirely from within.

Is there any evidence for this view? Wilder Penfield observed that, by electrically stimulating the temporal lobes, it was possible to evoke a memory so vivid and real that patients believed they were reliving an experience. Stimulating particular parts of the cortex evoked visual and auditory sensations along with emotions and feelings. Clearly, these experiences were evoked from within.

We know that animals can learn and remember things. What is it, then, about our memories that is special? Patient HM, who lost the ability to form long-term declarative memories when his hippocampi were removed, may provide a clue. He could still learn new skills and tasks, but he had no conscious memory of them. He learned things but had no awareness that he had learned them, and he even denied that he had done a learned task before. Is this what is special about our declarative memories—that when we elicit them we become aware of them? Are animals aware of what they have experienced? We really don't know.

The challenge before us is to understand these phenomena—sensory perception, memory and learning, language, planning and programming, and ultimately consciousness—in terms of neural mechanisms. At the moment this seems a daunting task, but we are getting glimpses of how neural activity can relate to complex behaviors. A nice example is working memory, which allows us to keep something in mind until we act on it. The classic example is looking up a phone number and remembering it until we dial the phone.

How neuronal activity participates in such phenomena is illustrated by recording neurons in the prefrontal area of the cortex, as discussed in Chapter 11. Certain neurons in the prefrontal cortex become active during the time when the monkey is remembering where a target is. The neurons are inactive before and after the completion of the task, but they are active continuously during the delay between the brief presentation of the target and the task they must undertake. If the delay is excessively long, the neurons gradually stop firing and the monkey is then unable to perform the task. If during the normal delay period a monkey becomes distracted, the neurons will also often stop firing, and again the animal cannot do the task reliably. (The same happens to us when we are distracted while trying to remember a phone number.) These neurons, then, are critical for the successful completion of the task. Their activity links the sensory stimulus and the eventual behavior.

Another tantalizing finding, by Giacomo Rizzolatti and his colleagues at the University of Parma, Italy, may relate to neural mechanisms underlying complex behaviors and even consciousness: the discovery of *mirror neurons* in the premotor cortex. Chapter 10 discusses how the premotor cortex has been implicated in the planning and programming of motor activity. If a human subject is asked to think about how to make a complex motor act but not to make the movements appropriate for the act, enhanced activity is observed by functional imaging in an area of the premotor cortex, but not in the area of the motor cortex that would initiate the appropriate movements.

Rizzolatti and colleagues found that there are neurons, called mirror neurons, in the premotor areas that become active when a monkey observes a motor act of another monkey. The observing monkey does not mimic the act of the other monkey, but when these neurons fire, the monkey presumably now knows what the other animal is doing, what its intentions are. There is indirect evidence (by imaging) that a mirror system exists in humans.

Consciousness is often described as a state of awareness, both self-awareness and awareness of the intentions of others. Mirror neurons can provide understanding of the intentions of others. Interestingly, understanding someone's intentions is deficient in those suffering from autism, and so it has been proposed that a defect in the mirror neuron system could be involved in the disorder. Whether this is so is not yet known, or whether similar neurons elsewhere in the brain are active in response to the intentions of others.

We are just at the beginning of understanding how neural activity underlies complex behavioral tasks and how they might relate to consciousness. The working memory example described above does not tell us how neural activity keeps the location of the target in the monkey's mind. What is known is that a neuron codes for a specific location in the visual field. That is, some neurons will become active only when the target appears at the three-o'clock position in the visual field, others when it appears at the twelve-o'clock position, and so forth. Thus, the neurons are coding for visual field position in addition to the delay.

The recording of neurons while animals, particularly monkeys, are awake and behaving will clearly tell us an enormous amount, and the techniques necessary to do this have now been quite well worked out, but there are challenges. For example, the animals must not feel discomfort, and considerable effort is expended to ensure this. Further, it is impossible to carry out such experiments in humans; the neural basis of human consciousness seems beyond our experimental reach for the time being. Nevertheless, the noninvasive imaging techniques described in Chapter 10 promise to reveal much about neural function in humans. Noninvasive techniques may well enable us eventually to record the electrical activity of specific neurons, or at least groups of neurons, in the human brain.

The Future

Where will we be in 50 or 100 years in terms of understanding brain mechanisms? No one knows, of course, but some speculations are possible. Certainly our increased understanding of brain mechanisms will have enormous impacts on how humans conduct their affairs. Already this is the case, illustrated by two examples: eyewitness reports, and the placebo effect. That memories are often reconstructive, depending on past and present events, experiences, and biases, clearly has legal implications. We now recognize that distortion of memories is a normal phenomenon. Memories are made logical and coherent by our brains—and they are affected by emotion and traumas. That eyewitnesses remember events quite differently is inevitable and does not mean that someone is being untruthful. The perception of an event varies among individuals, for similar reasons, so discovering the "truth" about an event is a challenge, and perhaps an impossibility in some cases.

The recognition that the placebo effect relieving pain can be explained by the release of endogenous opiate-like substances within the brain raises the intriguing possibility that a variety of chemicals contained within neurons can profoundly affect and alter the way we sense things or feel. The range of effects such substances might exert is conjecture at present, but, for instance, that there are receptors in the brain activated by marijuana suggests there are endogenous molecules in the brain that interact with these receptors. Under what conditions are these molecules released, and what are their effects? These are a few of the questions that undoubtedly will be answered in the next few years.

That the brain may contain such molecules as its own opiate-like substances, the enkephalins, is influencing our notions of how we test for the effectiveness of new drugs and of how psychotherapy works. When a subject is given a test drug and told that it may help him, is the effect subsequently noted due to the drug or to an endogenous molecule released in the brain? And is the ultimate effect of psychotherapy to induce or alter the release of certain neuroactive substances within the brain? Clearly, mind-body interactions and influences are viewed quite differently today than they were just a few years ago. And it is safe to predict that psychiatry, viewed often as a lagging discipline in medicine, will be the major beneficiary of advances in understanding brain mechanisms.

Is there a Rosetta Stone for neuroscience, which if deciphered would transform the field as did the elucidation of the structure of DNA for molecular biology? Not that we are aware. But no one imagined that understanding the structure of DNA would give us such marvelous insights into the way genetic material replicates and how protein structure is coded. A discovery in neuroscience tomorrow might provide as powerful an impact on our field and on our understanding of how the brain works—we simply don't know where to look. As noted in the preface, new powerful techniques are being developed to map the brain down to the synapse level, to record from hundreds of neurons simultaneously in awake-behaving animals, as well as new and better ways to analyze human brain activity noninvasively.

Artificial Intelligence

Much has been said and written recently about artificial intelligence and whether computers will soon be the equivalent of the human brain. Clearly computers and robots with the automation computers provide have revolutionized the way we live, how we do things, and how we make things, but will computers ever have minds that will compete with or surpass the human brain? We do not know, of course, but at the moment, we are a very long way from this happening, and serious reservations can be raised as to whether this will ever happen.

Increasingly, it is claimed that computers are mimicking the structure of the brain and that their design is based on "neural networks." But we still know so little of how neural circuits underlying higher brain function are structurally organized and how they work that using the term *neural networks* to describe how computers are designed is very much an exaggeration and even misleading. Perhaps when we do know more of how neural circuits are structured and how they function, we will be able to design computers that do indeed approximate the brain, but this is not the case today.

The BRAIN Initiative, announced by then President Obama in 2013, has as its goal a detailed analysis of the human brain both anatomically and physiologically, but this quest is still in its infancy. And as was the case with the other great biomedical initiatives undertaken in the last 50–60 years, I suspect new and fundamental discoveries will be made along the way that will alter significantly our views and understanding of how the brain functions. For example, when the war on cancer was initiated in the mid-1960s, it was to eradicate cancer. Have we succeeded? No, but we have learned an enormous amount about the disease and have had some significant successes in combating a number of cancers. When the war began, it was generally believed cancer was one disease, but today we know cancer is many diseases, caused by many factors. There is no magic bullet that can prevent or cure all cancers. This much we do know. Virtually every cancer type requires understanding on its own.

Next was the Genome Initiative in the 1990s, to analyze the human genome, which it was hoped would enable us to deal with all the genetic diseases that mankind suffers. It was then believed that one gene codes for one protein and the guess was that we had about 100,000 genes in our genome. Well, how many genes do we have, now that we have succeeded in analyzing the human genome? About 20,000. How can this be? That is not enough code for all the proteins we have. The answer is that one gene can code for more than one protein through a process called alternative splicing. Indeed, one gene, it has been estimated, could code for as many as 30,000 different proteins, although this is the exception and not the rule. But this was totally unexpected and has made the analysis of the human genome and its protein products much more complicated.

Such is likely to be the case with the BRAIN Initiative. For example, critical aspects of brain function such as consciousness remain very much of a mystery, and even defining consciousness defies us still. Is it realistic to suggest that computers will ever have consciousness as we humans do when we don't really know what consciousness is?

A final comment concerns the idea of artificial intelligence. Albert Einstein remarked once that "the true sign of intelligence is not knowledge but imagination." We can make computers knowledgeable, but can we imbue them with imagination and creativity? We do not know, and would we even want to do this? The latter is another question to be considered, but first we need to go much, much further in understanding the brain and how it functions before we can say computers have achieved real artificial intelligence.

Glossary

acetylcholine Neurotransmitter released at the neuromuscular junction; also released from certain synapses in the brain, where it can have either neurotransmitter or neuromodulatory effects, and from parasympathetic nervous system neurons.

action potentials The transient, all-or-nothing electrical signals that travel down axons carrying the output information of neurons.

adaptation The decline in responsiveness of sensory receptors to a sustained stimulus as a function of time.

adenosine triphosphate *(ATP)* An energy-rich molecule that powers biochemical reactions in cells.

adenylate cyclase The enzyme that converts ATP to cAMP.

agnosia (visual) The inability to recognize objects

amacrine cell One class of inner retinal neuron. Many amacrine cells are movement sensitive.

amblyopia The loss of visual acuity because of visual (form) deprivation of an eye or a crossed eye in a young animal or human.

amino acids The molecules that when strung together form proteins.

amygdala A region in the forebrain involved in integrating and coordinating emotional behaviors.

Aplysia californica An invertebrate used especially for studies of elementary forms of learning and memory.

area V1 The primary visual area of the brain, where visual information is first processed in the occipital cortex.

areas V2–V8 Areas in the occipital cortex that, along with area V1, are concerned with processing visual information.

association areas Regions of the cerebral cortex concerned with higher levels of processing.

association neurons Cells that mediate interactions between neurons.

autonomic nervous system The part of the nervous system that regulates our internal organs. Much of the regulation is involuntary and mediated by two opposing subdivisions, the sympathetic and parasympathetic systems.

autoreceptors Receptors found on synaptic terminals that are activated by the substances released by the terminals.

axon Thin cellular branch that extends from a neuron to contact another neuron or a muscle cell. Axons carry the output message from a neuron via action potentials.

axon terminals Enlargements of an axon near its site of termination where synapses are typically made.

axonal transport The special mechanism by which substances are moved rapidly down axons.

basal ganglia Five brain nuclei found in the forebrain that are concerned with the initiation and execution of movements.

basilar membrane The membrane in the organ of Corti in which the hair cells are imbedded.

bipolar cell A class of retinal neuron that carries the visual signal from the outer to inner retina.

birdsong The songs birds sing.

brain imaging (PET and fMRI) The imaging of increases of blood flow that occur when neurons in a brain area are active.

bride of sevenless (boss) A mutation in the fruit fly that prevents the R7 photoreceptor from developing.

Broca's area An area (usually found in the left frontal lobe of the cerebral cortex) critical for the production of language.

calcium (Ca^{2+}) A positively charged ion important in synaptic transmission and that also serves as a second messenger.

calmodulin A protein that binds calcium ions. Activated calmodulin activates a specific kinase termed CaM kinase (and other kinases as well) (CaMK, Ca^{2+}/calmodulin-dependent protein kinase).

catecholamine A class of monoamine derived from the amino acid tyrosine. Dopamine, norepinephrine, and epinephrine are examples.

central nervous system The part of the nervous system that comprises the brain and spinal cord.

cerebellum A prominent hindbrain structure important for coordinating and integrating motor activity.

cerebral cortex A 2-millimeter-thick layer of cells that covers the forebrain. Highly infolded in man, the cortex is divided into two hemispheres , which are further subdivided into four lobes, frontal, parietal, occipital, and temporal.

cerebrum A collective term for the cerebral cortex, basal ganglia, and associated structures.

channels Membrane proteins that allow ions to cross the cell membrane. Channels are usually closed until activated by a specific stimulus.

chloride (Cl⁻) A negatively charged ion primarily involved in the inhibition of neurons.

circadian rhythm Endogenous rhythm that regulates various bodily functions depending on time of day.

cochlea A coiled structure containing three fluid-filled chambers as well as the sensory receptors (hair cells) and accessory structures that underlie audition (hearing).

complex cells Neurons recorded in the primary visual area of the cortex that respond best to oriented bars of light or dark moving at right angles to the bars' orientation .

cone photoreceptors The photoreceptors responsible for color vision. Three types of cones exist in the human retina, sensitive to red, green, or blue light.

corpus callosum A thick band of axons found in the middle of the brain that carries information from one side of the brain to the other.

cortical columns Columns of neurons that run through and across the cortex that share similar properties, such as orientation or ocular dominance columns in area V1 of the visual cortex.

cortisol A steroid hormone released from the adrenal glands especially during stress.

cranial nerves Twelve nerves that enter the brain directly. Ten carry sensory and/or motor information related to the head, and two innervate the internal organs.

critical period The period of time during development when an animal is particularly sensitive to environmental conditions.

curare A drug that paralyzes muscles by blocking the acetylcholine receptors found on muscle cells.

current A measure of the flow of electrons through a wire, or ions across a membrane per unit of time.

cyclic adenosine monophosphate (cAMP) A second-messenger

molecule formed by the enzyme adenylate cyclase from adenosine triphosphate (ATP).

cytoplasm The substance inside cells exclusive of the nucleus.

dark adaptation The time required for the eye to regain full sensitivity after light exposure.

declarative memory The memories of facts or events.

dendrites Bushy, branch-like structures that extend from the cell body of a neuron and receive the synaptic input to the cell.

direction-selective cell A cell in the visual system that responds selectively to a spot or bar of light or dark moving in a particular direction across the retina.

disinhibition The inhibition of an inhibiting neuron, resulting in a partial relief of inhibition.

disparity-tuned cells Neurons that respond only to stimuli precisely positioned within their receptive fields, thought to be critical for depth perception.

dopamine A neuromodulator released from brain synapses that has been associated with two diseases, Parkinson's disease and schizophrenia.

eccentric cell Second-order neuron in the horseshoe crab eye.

ectoderm Cells on the outer surface of the embryo that become skin.

electrical synapse A junction where ions flow directly from one cell to the next.

electrons The small negatively charged particles that surround the protons in an atom.

endoderm Cells lining the inside of an embryo that form the gut and other internal organs.

enkephalins Small peptides released from synapses in the brain that have opiate-like effects.

epilepsy Seizures caused by diseased or damaged cells in the brain.

excitatory synapse A synapse that excites a neuron or muscle cell.

forebrain The most distal part of the brain, consisting principally of the thalamus, hypothalamus, basal ganglia, and cerebral cortex.

fovea The central region of the eye that mediates high-acuity vision.

frontal lobe The most anterior portion of the cerebral cortex, concerned primarily with movement, planning, programming, and smell.

functional magnetic resonance imaging (fMRI) A technique that measures increases in blood flow in an active region of the brain.

γ-aminobutyric acid (GABA) An inhibitory neurotransmitter in the brain.

ganglia Groups of nerve cells that usually serve a particular function; typically used to describe groups of neurons outside the central nervous system or in invertebrate nervous systems, but there are exceptions, such as the basal ganglia in the forebrain.

ganglion cells The third-order cells in the retina whose axons form the optic nerve; also, cells in a ganglion.

generator potential A collective term for excitatory synaptic and receptor potentials that lead to the generation of action potentials.

germinal zone Where neurons and glial cells are generated in the embryo; initially on the inner surface of the neural tube.

glia Supporting cells in the brain that help maintain neurons, prune processes, regulate the environment, and form the myelin around axons.

glomeruli Structures in the olfactory bulb that receive input from odorant receptors of a specific type.

glutamate An amino acid that serves as the major excitatory neurotransmitter in the brain.

glycine An amino acid that serves as an inhibitory neurotransmitter in the brain.

Golgi method A silver-staining method discovered by Golgi, an Italian histologist, but used prominently by Santiago Ramón y Cajal, a Spaniard, to elucidate neuronal structures.

G-protein A protein activated by postsynaptic membrane receptors, usually linked to an enzyme that makes a second-messenger molecule.

gray matter Those regions of the brain and spinal cord where neuronal cell bodies and dendrites are abundant.

grid cells Neurons found in the cortex near the hippocampus that are active when an animal is in several locations (place fields), which are arranged in a triangular pattern. Grid cells and place cells in the hippocampus are thought to interact, but how is not clear.

growth cone The specialized end of a growing axon.

growth factors Small proteins important for cell growth, differentiation, and survival.

guidepost neurons Specialized cells found in the developing brain that guide axonal growth.

habituation The decrease in the strength of a response following repeated elicitations of the response.

hair cells The sensory receptors for audition (hearing). They respond to bending of the fine hairs that project outward from their apical surface.

hemisphere (cortical) Half of the cerebral cortex. The two cortical hemispheres are each subdivided into four lobes, frontal, parietal, occipital, and temporal.

hindbrain The lowermost part of the brain, emerging from the spinal cord; consists mainly of the medulla, pons, and cerebellum.

hippocampus A region of the brain found under the temporal lobes important for establishing long-term memories.

histology The microscopic study of tissues.

horizontal cell An outer retinal cell class that mediates lateral inhibition between photoreceptors and bipolar cells.

horseshoe crab An invertebrate, *Limulus polyphemus,* useful for studies of visual mechanisms.

Huntington's disease An inherited disease of the basal ganglia that causes movement dysfunction.

hypercolumn (module) The basic module of the visual cortex. A piece of cortex $1 \times 1 \times 2$ millimeters that contains all the machinery needed to analyze a bit of visual space.

hypothalamus A forebrain region that contains nuclei concerned with basic acts and drives such as eating, drinking, and sexual activity. The hypothalamus also regulates the release of pituitary gland hormones and the autonomic nervous system and plays an important role in emotional behavior.

indoleamine A class of monoamine derived from the amino acid tryptophan. Serotonin is an example.

inhibitory synapse A synapse that inhibits neurons.

interneurons A general term for all neurons that are found between sensory neurons and motor neurons.

invertebrates Animals such as insects, crabs, flies, and mollusks that do not have a backbone.

ions Atoms that are charged; that is, have gained electrons and are thus negatively charged or have lost electrons and are thus positively charged.

kinases Enzymes that add phosphate groups to proteins, thereby altering their function.

lateral geniculate nucleus (LGN) The nucleus in the thalamus that receives input from the eye and transmits the visual signal to the cerebral cortex.

lateral inhibition The reciprocal inhibition of one neuron by another, best characterized in the horseshoe crab eye.

L-dopa A precursor of dopamine useful for treating Parkinson's disease.

Limulus polyphemus The horseshoe crab.

long-term depression (LTD) A persistent decrease in synaptic potential amplitude induced in a neuron by a strong priming stimulus delivered to the neuron.

long-term memory Memories that last for long periods—weeks, months, or longer.

long-term potentiation (LTP) A persistent increase in synaptic potential amplitude induced in a neuron by a strong priming stimulus delivered to the neuron.

Mach bands Light and dark bands seen adjacent to dark and light borders, respectively, that enhance edge detection.

mechanoreceptor Sensory receptors that respond directly to deformation of membrane channels themselves or of the surrounding membrane.

medulla A hindbrain region that contains nuclei involved with vital body functions, including heart rate and respiration.

membrane (cell) The thin barrier surrounding a cell that keeps various substances in and other substances out; consists of a lipid bilayer in which are embedded various kinds of proteins, including channels, pumps, enzymes, and receptors.

mesoderm Cells between the ectoderm and endoderm in the embryo that develop into muscle, bone, and heart cells. In the early embryo, mesodermal cells induce overlying ectodermal cells to become neural plate cells.

messenger RNA The RNA that carries the code for a protein from the DNA in the nucleus to a ribosome in the cytoplasm where the protein is made.

midbrain That part of the brain between the hindbrain and forebrain.

mirror neurons Cells in the frontal lobes that become active when an animal observes another animal carrying out a specific task.

mitochondria Structures found in cells that provide the energy-rich molecules (i.e., adenosine triphosphate) that power cells.

mitral cells The major output neuron of a glomerulus in the olfactory bulb.

monoamine A type of substance released at synapses that functions mainly as a neuromodulator.

monosynaptic reflex A simple reflex circuit consisting of a sensory neuron impinging directly on a motor neuron.

motor output The activity of motor neurons that results in muscle movements.

multiple sclerosis (MS) A disease of the myelin that surrounds axons.

myasthenia gravis A disease of the neuromuscular junction.

myelin An insulating layer of membrane formed around axons by glial cells.

nerve growth factor (NGF) The first-identified and best-studied growth factor found in the brain.

neural crest Cells derived from the neural plate that form much of the peripheral nervous system.

neural plate Cells found on the dorsal side of an embryo that form the nervous system.

neural tube An early stage in the development of the brain formed by the infolding of neural plate cells.

neuromodulator Substance released at a synapse that causes biochemical changes in a neuron.

neurons Cells in the brain involved in the reception, integration, and transmission of signals.

neuropeptides Small proteins (peptides) released at synapses that act mainly as neuromodulators.

neurotransmitter Substance released at a synapse that causes fast electrical excitation or inhibition of a neuron.

night blindness Loss of visual sensitivity at night; can be caused by a deficiency of vitamin A or certain inherited eye diseases.

NMDA channels Channels that respond to glutamate and allow Na^+ and Ca^{2+} to enter a neuron.

nodes Those regions along an axon where the myelin is interrupted and action potentials are generated.

norepinephrine A catecholamine released at certain synapses in the brain and also from sympathetic nervous system neurons.

nucleic acid (DNA) The genetic material found in the nucleus of a cell that codes for the proteins made by the cell.

nucleus A cluster of neurons in the brain that generally serves a particular function; also, the structure within cells that contains the genetic material (DNA).

nucleus accumbens A nucleus found in the basal region of the forebrain involved in reinforcing behaviors such as drinking when thirsty or eating when hungry. This nucleus has also been implicated in addictive behaviors.

occipital lobe The most posterior portion of the cerebral cortex, concerned with visual processing.

ocular dominance The phenomenon that results in most binocular cells in the cortex being driven more strongly by one eye than the other.

OFF-center cell A neuron whose activity decreases in response to stimulation of its receptive field center. Usually at the cessation of the stimulus such cells become active for a short period.

olfaction Smell.

oligodendrocytes Glial cells in the brain and spinal cord (central nervous system) that form myelin.

ommatidium The photoreceptive unit found in the eyes of invertebrates.

ON-center cell A neuron whose activity increases in response to stimulation of its receptive field center.

ON-OFF cell A neuron whose activity increases at the onset of a stimulus and again at the cessation of stimulation.

orbitofrontal cortex An area found in the lower part of the frontal lobes, important for the expression of emotional behaviors.

organ of Corti The structure in the inner ear that results in hearing, consisting of hair cells and the basilar and tectorial membranes.

Pacinian corpuscle A sensory mechanoreceptor receptor that responds to touch, pressure and vibration.

parallel processing The simultaneous processing of information along separate neural pathways.

parasympathetic nervous system The division of the autonomic nervous system that promotes "rest and digest" behaviors.

parietal lobe That region of the cerebral cortex between the frontal and occipital lobes concerned primarily with somatosensory information processing.

Parkinson's disease A disease of the motor system caused by a deficiency of dopamine in the basal ganglia. Patients with the disease typically develop a tremor and have difficulty initiating movements.

peptide A small protein.

peripheral nervous system Parts of the nervous system outside of the brain and spinal cord.

phosphodiesterase (PDE) An enzyme that breaks phosphate bonds.

phosphorylation The addition of a phosphate group to a protein; serves to modify the properties of the protein.

pioneer axons Axons that form early in development and provide a path for other axons to follow.

pituitary gland A gland found at the base of the brain that releases a variety of hormones into the bloodstream.

place cells Neurons present in the hippocampus that signal when an animal is in a specific location

placebo An inert substance that can cause physiological effects under certain circumstances.

pleasure centers Regions of the brain that when stimulated appear to give pleasure to an animal. These regions appear to be related to the reinforcing and reward systems of the brain.

pons A hindbrain structure that relays information from the cortex to the cerebellum.

positron emission tomography (PET) A method for detecting increases in activity of a part of the brain.

postsynaptic Pertaining to structures or processes downstream of a synapse, for example, postsynaptic neuron, postsynaptic potential.

postsynaptic membrane That region of a cell membrane specialized to receive synaptic input.

potassium (K⁺) A positively charged ion primarily involved in establishing the resting potential.

potential Another term for voltage.

prefrontal lobotomy A surgical procedure that severs the connections between the orbitofrontal cortex and the rest of the brain.

premotor areas Regions in the frontal lobes thought to be important for the planning and programming of motor movements.

presynaptic Pertaining to structures upstream of a synapse, for example, presynaptic neuron, presynaptic terminal.

presynaptic synapse A synapse made onto a synaptic terminal.

primary motor area The region of the cerebral cortex where fine movements are initiated, found in the frontal lobes adjacent to the central sulcus.

primary sensory areas Regions where sensory information is first processed in the cerebral cortex.

procedural memory The memory of a motor skill such as riding a bicycle.

proprioceptive information Sensory information from muscles, joints, and tendons of which we are not aware—it does not reach our consciousness.

prosopagnosia The inability to recognize faces.

protein A chain of amino acids folded in complex ways that enable the molecule to carry out its prescribed function.

protons The positively charged particles found in the center of an atom.

pump A membrane protein that moves ions across the membrane of a cell. Pumps require energy to function.

Purkinje cell A large neuron found in the cerebellum.

pyramidal cell A prominent neuron found in all areas of the cerebral cortex.

radial glial cell A specialized glial cell found in the developing brain along whose processes precursor neurons travel to find their appropriate position.

rapid eye movement (REM) sleep An active phase of sleep during which most dreaming occurs.

receptive field That area of the retina that when stimulated causes a retinal cell to alter its activity.

receptor Membrane protein found in postsynaptic membranes that is usually linked to intracellular enzyme systems; also, a cell that responds to specific sensory stimuli, for example, a photoreceptor.

receptor potential The voltage change elicited in a sensory cell or neuron following the presentation of a specific sensory stimulus to the cell or neuron.

reflex An involuntary motor response in response to a specific stimulus.

resting potential The voltage across a cell membrane in the absence of any stimulus to the cell.

reticular formation Neurons found throughout the medulla that extend widely in the brain and are important for regulating states of arousal and levels of consciousness.

retinal The aldehyde form of vitamin A that, when combined with a specific protein, forms a visual pigment molecule.

retinitis pigmentosa An inherited disease that causes a slow degeneration of the photoreceptors.

rhodopsin The visual pigment of rods.

ribosomes Particles found in cells that are responsible for making proteins.

rod photoreceptors The photoreceptors responsible for dim-light vision.

schizophrenia A severe mental disease characterized by thought and mood disorders, hallucinations, and so forth.

Schwann cells Glial cells of the peripheral nervous system that form the myelin around axons.

second messenger A small molecule synthesized in a cell in response to a neuromodulator (the first messenger).

sensitization The increase in the strength of a response following the presentation of an adverse stimulus to an animal.

serotonin A substance released at synapses that most often acts as a neuromodulator. Decreased levels of serotonin in the brain have been linked to depression.

sevenless A genetic mutation in the fruit fly that prevents the R7 photo-receptor from developing.

short-term memory The initial storage of memories that lasts for a few minutes or so. Short-term memories are labile and easily disrupted.

simple cells Neurons recorded in the primary visual area of the cortex that respond best to oriented bars of light or edges projected onto the retina.

sodium (Na⁺) A positively charged ion involved in the generation of action potentials and in the excitation of neurons and sensory cells.

somatosensory Pertaining to sensory information coming from the skin and deeper tissues of the limbs and trunk, such as touch, pressure, temperature, and pain.

squid An invertebrate (a mollusk) that has giant axons.

stem cells Undifferentiated cells that can proliferate and become any type of cell.

sulcus A prominent and deep infolding of the cerebral cortex.

supplementary motor area A premotor area involved in the planning and programming of motor movements.

sympathetic nervous system The division of the autonomic nervous system that mediates "fight or flight" reactions.

synapse The site of functional contact between two neurons or a neuron and muscle cell.

synaptic potential The voltage change produced in a neuron following the activation of a synapse impinging on the cell.

synaptic terminal A site where a synapse is usually made.

synaptic vesicles Small vesicles found at synapses that contain the chemicals released at the synapse.

tectorial membrane The membrane in the organ of Corti that causes the bending of the hairs on the hair cells, resulting in cell excitation.

tectum A midbrain structure, especially prominent in nonmammalian species, which integrates sensory inputs and initiates motor outputs.

temporal lobe The lateral-most part of the cerebral cortex, concerned with hearing and memory.

tetrodotoxin An antagonist that blocks voltage-gated Na^+ channels.

thalamus A forebrain region that relays sensory information to the cerebral cortex.

topographic Pertaining to the orderly projection of axons from one region of the brain to another.

transcription factor A protein that interacts with DNA to turn on or off the expression of a gene.

transducin A G-protein activated by a visual pigment and that activates phosphodiesterase.

transmagnetic stimulation Stimulation or inhibition of neural activity by a magnetic beam.

transporter A membrane pump that transports substances released at a synapse back into the synaptic terminal, thereby terminating the activity of the substances.

tricyclics Drugs that raise the levels of monoamines in the brain by inhibiting their reuptake into synaptic terminals.

tryptophan The amino acid from which indoleamines such as serotonin are derived.

tyrosine The amino acid from which catecholamines such as dopamine and norepinephrine are derived.

visual area Area in the occipital cortex involved with processing of an image.

visual pigment Molecule in photoreceptors that absorb light and leads to the excitation of the cell.

voltage A measure of the electrical charge difference between two points; in neurons, the charge difference across the cell membrane (i.e., membrane voltage or potential).

voluntary motor system The part of the nervous system that controls the muscles of the limbs, body, and head. The control is mainly voluntary.

Wernicke's area An area in the left temporal lobe concerned with the comprehension of language, reading and writing.

what pathway The ventral visual pathway that is involved in object recognition.

where pathway The dorsal visual pathway that is involved in spatial visual tasks.

white matter Regions of the brain and spinal cord where there are abundant myelinated axons. The myelin gives the tissue its whitish appearance.

working memory A memory maintained for a short time to enable a specific task to be accomplished. An example is remembering a phone number until it is dialed.

X-chromosome The so-called sex chromosome. Males have just one; but females have two.

Further Reading

A collection of books on various brain-related topics. All are accessible, and most are a good read.

Andreasen, Nancy C. (2004). *Brave new brain: Conquering mental illness in the era of the genome* (paperback ed.). New York, NY: Oxford University Press.

Bear, Mark, Connors, Barry, and Paradiso, Michael. (2016). *Neuroscience: Exploring the brain* (4th ed.). Philadelphia, PA: Wolters Kluwer.

Berkowitz, Ari. (2016). *Governing behavior: How nerve cell dictatorships and democracies control everything we do*. Cambridge, MA: Harvard University Press.

Cajal, Santiago Ramón y. (1989). *Recollections of my life*. Cambridge, MA: MIT Press.

Crick, Francis. (1994). *The astonishing hypothesis: The scientific search for the soul*. New York, NY: Charles Scribner's Sons.

Damasio, Antonio R. (1994). *Descartes' error: Emotion, reason, and the human brain*. New York, NY: Grosset/Putnam.

Dennett, Daniel C. (1992). *Consciousness explained*. Boston, MA: Little, Brown.

Doidge, Norman. (2007). *The brain that changes itself: Stories of personal triumph from the frontiers of brain science*. New York, NY: Penguin.

Dowling, John E., and Dowling Joseph L. (2016). *Vision: How it works and what can go wrong*. Cambridge, MA: MIT Press.

Dunbar, Robin, Barrett, Louise, and Lycett, John. (2005). *Evolutionary psychology: A beginner's guide* (paperback ed.). Oxford, UK: One World.

Herman, Dorothy. (1998). *Helen Keller: A life* (paperback ed.). Chicago, IL: University of Chicago Press.

Hobson, J. Allan. (1989). *The dreaming brain*. New York, NY: Basic Books.

Hubel, David H. (1988). *Eye, brain, and vision*. New York, NY: Freeman.

Kaku, Michio. (2014). *The future of the mind: The scientific quest to understand, enhance, and empower the mind*. New York, NY: Anchor Books.

Kandel, Eric. (2007). *In search of memory: The emergence of a new science of mind* (paperback ed.). New York, NY: Norton.

Koch, Christopher. (2012). *Consciousness: Confessions of a romantic reductionist*. Cambridge, MA: MIT Press.

Kramer, Peter D. (1997). *Listening to prozac*. New York, NY: Penguin USA.

LeDoux, Joseph. (2002). *Synaptic self: How our brains become who we are*. New York, NY: Penguin.

Luria, A. R. (1968). *The mind of a mnemonist*. New York, NY: Basic Books (also available with new introduction in paperback from Harvard University Press, Cambridge, 1987)

Morange, Michel. (2001). *The misunderstood gene*. Cambridge, MA: Harvard University Press.

Pinker, Steven. (1994). *The language instinct*. New York, NY: Morrow.

Pollen, Daniel A. (1993). *Hannah's heirs: The quest for the genetic origins of Alzheimer's disease*. New York, NY: Oxford University Press.

Posner, Michael I., and Raichle, Marcus E. (1994). *Images of mind*. New York, NY: Scientific American Library.

Sacks, Oliver. (1995). *An anthropologist on Mars*. New York, NY: Knopf.

Sacks, Oliver. (1987). *The man who mistook his wife for a hat and other clinical tales*. New York, NY: Harper Collins.

Schacter, Daniel L. (1996). *Searching for memory: The brain, the mind and the past*. New York, NY: Basic Books.

Snyder, Solomon H. (1989). *Brainstorming: The science and politics of opiate research*. Cambridge, MA: Harvard University Press.

Thompson, Richard F., and Madigan, Stephen A. (2007). *Memory: The key to consciousness* (paperback ed.). Princeton, NJ: Princeton University Press.

Valenstein, Elliot S. (1986). *Great and desperate cures: The rise and decline of psychosurgery and other radical treatments for mental illness*. New York, NY: Basic Books.

Watson, Peter. (2016). *Convergence: The idea at the heart of science*. New York, NY: Simon & Schuster.

Zeki, Semir. (1993). *A vision of the brain*. Oxford, UK: Blackwell.

Index

Note: Italicized page locators refer to illustrations.

About the Author

John E. Dowling is the Gordon and Llura Gund Research Professor of Neuroscience at Harvard University. Now semi-retired, he taught the introductory course on behavioral neuroscience at Harvard College for over 30 years. He now teaches a freshman seminar titled "The Amazing Brain." Long interested in the retina of the eye as an accessible and approachable part of the brain, he has published more than 270 research papers, and five books, two of which have been republished as revised editions. He is an elected member of the National Academy of Sciences, the American Academy of Arts and Sciences, and the American Philosophical Society. He has received several awards for his research on the visual system including the Friedenwald Medal from the Association for Research in Vision and Ophthalmology, the Von Sallman Prize from the International Society for Eye Research, the Helen Keller Prize for Vision Research, the Prentice Medal from the American Academy of Optometry, and Glenn A. Fry Medal in Physiological Optics. He and his wife Judy live in Boston.